LEGENDS, LIES & CHERISHED MYTHS OF WORLD HISTORY

Also by Richard Shenkman

"I Love Paul Revere, Whether He Rode or Not"

Legends, Lies & Cherished Myths of American History

One-Night Stands with American History (with Kurt Reiger)

LEGENDS, LIES & CHERISHED MYTHS OF WORLD HISTORY

RICHARD SHENKMAN

Illustrations by George J. McKeon

HarperPerennial
A Division of HarperCollins*Publishers*

A hardcover edition of this book was published in 1993 by HarperCollins Publishers.

HarperCollins books may be purchased for educational, business, or sales promotional use. For information, please write: Special Markets Department, HarperCollins Publishers, Inc., 10 East 53rd Street, New York, NY 10022.

First HarperPerennial edition published 1994.

Designed by George J. McKeon

The Library of Congress has catalogued the hardcover edition as follows:
Shenkman, Richard.
Legends, lies & cherished myths of world history / Richard Shenkman.—1st ed.
p. cm.
Includes bibliographical references and index.
ISBN 0-06-016803-X (cloth)
1. History—Errors, inventions, etc. 2. History—Humor.
I. Title. II. Title: Legends, lies and cherished myths of world history.
D10.S52 1993
902.07—dc20 92-56210

ISBN 0-06-092255-9 (pbk.)

94 95 96 97 98 ❖/HC 10 9 8 7 6 5 4 3 2 1

For my mother
Phyllis Shenkman
who keeps growing

CONTENTS

PART 6: "THIS SCEPTER'D ISLE"

(Or: British history the way it should have been taught)

PART 7: LET THEM EAT BRIOCHE!

(Or: French history for beginners)

PART 8: LIKEABLE (AND NOT-SO-LIKEABLE) FAMOUS PEOPLE

(Or: If you learned it in school, it can't be true)

SOME THINGS YOU SHOULD KNOW BEFORE READING THIS BOOK

We Americans, I have discovered, do not just get our own history wrong. We get everybody else's wrong as well.

Think Nero fiddled while Rome burned? Think Catherine the Great was Russian? Think King Arthur lived in a castle? (Think there really was a King Arthur?) Think Cleopatra was beautiful? Americans think these things are true, but they aren't.

Take almost any famous event of world history, from the Trojan War to World War II. The version we learned in school or at the movies was often cockeyed or bogus.

The plain fact is we have been flimflammed: We have been conned into believing that the pagan barbarians who overran the Roman Empire held civilization in contempt. We have swallowed the old line that English liberty can be traced to the signing of Magna Carta. And we have been duped into believing that the English endured the Blitz with a stiff upper lip.

These are the facts: Most barbarian tribes converted to Christianity, intermarried with the Roman elite, and joined the imperial army to defend the empire from its enemies. Magna Carta gave new rights only to England's powerful barons. And during the Blitz the English complained and were bitter; and many turned to crime.

Much of our history is topsy-turvy. Captain Bligh, a genuine hero, is made out to be a sadistic menace. Edward VIII, an open Nazi sympathizer, is remembered as the noble king who gave up his crown for the love of a woman. Hirohito, an ally of the Japanese militarists, is thought of as the shy marine biologist in glasses who hated war.

It would be going too far to say that our heads are completely filled with lies. It is simply that in many cases history is written by the victors and is filtered through the prism of their prejudices. Take the Spanish Inquisition. Why is it thought to have been one of the lowest, meanest, most reprehensible forums of injustice in human history? Not because it was, but because English Protestants wrote the history books.

Why are the Dark Ages regarded as dark? Because the Renaissance humanists hoped to leave the impression that they had rescued the world from gross ignorance.

Why did historians for so long ignore sex and history? They didn't use to, but Victorian historians took the sex out.

Why is Richard III depicted as a mean hunchback with a withered arm? Because Shakespeare wanted to make Richard's Tudor successors look better by comparison.

I'm asked a lot of times if it isn't a good thing that we have myths. Sure it is. The myths tell us who we are and what values we cherish, and every society has them. And if we didn't have them, some critic somewhere would be sure to say there's something wrong with us for not having myths like other people do.

But if everybody has myths, why bother debunking them? The answer is plain enough: we ought to know the truth about things.

The truth can be painful, but it must be faced. We need to know that Winston Churchill initially wanted to appease Hitler and that Franklin Roosevelt appeased Mussolini. We need to know that German P.O.W.s died by the thousands in American prisons at the end of World War II and that this information was concealed from the public. We need to know that footage in the old newsreels was often faked.

How do you know you can trust me to tell you the truth?

Actually, you shouldn't trust me. Indeed, you shouldn't trust anybody who writes history. We are all full of it. Despite the work of thousands of Ph.D.'s, truth in history is as difficult to ascertain today as it ever was. This is a fact. That's why this book is so valuable. For the author of this book (me) admits that what you have here is my version of the truth. It is the truth, the whole truth, and nothing but the truth—as I see it.

Truth, in short, is relative. It is in the eye of the beholder. But in saying this I am not saying there are no facts in history. There are. The Holocaust is a fact. The Americans who said in a recent poll that it's possible the Holocaust did not take place are wrong.*

Much of the stuff in this book, I know, sounds like I made it up. I didn't. The information is in buried the works cited in the source notes.

If the stories I tell seem crazy it is because, as my friend Bernard Weisberger says, life is crazy and people do damn fool things.

Some may think it's absurd to take on the history of the world. It is. But fortunately this book doesn't really cover all of world history, just the world history with which Americans are already familiar. Limiting the book in this way considerably narrows the areas that need to be dealt with.

What Americans mostly know about, of course, is European history, and of European history, what Americans mostly know is English history. There is a simple explanation for this. It was the descendants of the English who first decided what Americans should know about history. Naturally, they tended to favor their own kind.

---

* According to a 1993 poll conducted by the Roper organization, twenty-two percent of Americans believe it's possible the Holocaust never happened; another twelve percent said they did not know if it did.

PART 1

WAY BACK WHEN

TROJAN WAR

SOCRATES

ALEXANDER THE GREAT

HERODOTUS

CAESAR

CLEOPATRA

CALIGULA

NERO

THE FALL OF THE ROMAN EMPIRE

THE BARBARIANS

Nero

TROJAN WAR

The myth about the Trojan War is that there was one. There wasn't. At least there wasn't one that we know of. In the thousands of years that have elapsed since Homer's epic appeared, nobody has ever produced any evidence that the war he described took place. All the faithful have going for them is hope. (We don't even know if Homer was real. See below.)

That Troy once existed is true. Indeed, from archaeological evidence unearthed in the nineteenth and early twentieth centuries there would appear to have been at least nine Troys piled one atop the other (located in what is now Turkey). But there is no proof there was ever a war between Greece and Troy involving a beautiful queen named Helen, a big wooden horse, or a hero weakened by an Achilles' heel.

Presumably Greeks and Trojans fought each other at one time or another. After all, they were human. And there must have been some reason the Trojans built the huge walls surrounding their city. But there's no archaeological evidence that an army ever planted itself outside the walls of Troy, let alone a huge Greek army that is supposed to have numbered 110,000 soldiers.

Much of the story, at any rate, is patently implausible. That the war lasted ten years is inconceivable; army discipline never could have been maintained that long (no other war at the time is known to have lasted more than a few months). And nobody believes that the Greek soldiers camped out on the beach all those years, their Greek kings right along with them. The business about Helen—that she'd supposedly eloped with a Trojan prince and that the Greeks went to war to get her back—is attractive but unsubstantiated. Besides, it's unlikely she ever would have eloped. FitzRoy Raglan, an expert in world history, reported that he could find "no instance" in history "in which a queen has eloped with a foreign prince, or anybody else."

Anyway, nobody knows if Helen ever even lived. To be sure, tradition has it that the beauty whose face "launched a thousand ships" actually lived and actually served as queen. But tradition also has it that she was the daughter of Zeus and that she was "hatched from a swan's egg."

As for the story of the Trojan Horse, nothing substantiates it. Out of the thousands of objects that have turned up in repeated excavations of Troy, not one lends any credence to the existence of a big wooden horse.

Those who claim the story of Troy is true insist it doesn't matter if some of the details are implausible or unsupported. What counts are the plausible details. But by this method any poem could be found to be historically sound. Just because a poem includes a real person or two doesn't mean the poem is about a real event. Yet this is the kind of argument apologists for the Homeric epic have advanced.

Thucydides believed that the story of Troy was true. But Thucydides lived more than eight hundred years after the war supposedly occurred and was in no better position than we are to vouch for its accuracy. Probably he just wanted to believe it was true.

Homer has long been credited with the story but nobody knows who he was, where he lived, whether he really existed, or how he possibly could have come by reliable information about Troy's early history. If he lived it was in the eighth or ninth century B.C., some four centuries after the war he described was fought. Chances are we know more today about the real Troy than Homer would have.

It's possible, of course, that the story was handed down over the centuries largely intact. In the old days of oral tradition people had better memories than they do today. But why would the Greeks have bothered to celebrate a war with Troy when they neglected to recall so much else that happened in their past of far greater consequence?

What we are left with then is a poem written by a man who may not have lived concerning a war that probably never took place.[1]

SOCRATES

How did Socrates die? From the familiar depictions of the event it always looks as if he passed away peacefully. How did he actually die? He died a nasty, terrible, horrible death. After drinking his cup of hemlock, he went into convulsions, got nauseated, vomited, and then became paralyzed.

It was the great Plato who led people to think Socrates died in quiet dignity, but Plato, we now know, lied.

How do we know this? Because, after twenty-five centuries of research into every facet of Socrates' life, somebody one day finally thought to ask how it was that Socrates died a quiet death when everybody else who ever ingested a fatal dose of hemlock died in agonizing pain.

Speaking of Plato, how is it he was the one who chronicled the story of Socrates' death? Plato didn't even attend Socrates' death. Fourteen other disciples found the time to attend, but not dear old Plato.

Plato's excuse was that he was sick. But nobody believes him. You don't hear much about this, but historians think he stayed away from the death scene to deliberately distance himself from Socrates, who wasn't too popular a figure with the authorities in town just then.[2]

ALEXANDER THE GREAT

Alexander the Great was the first person in history to prove that killing lots of people is easy if you put your mind to it.

Killing ran in the family. His father, Philip II, demonstrated a talent for killing Greeks. His mother, Olympias, who worshiped snakes, had the young children of one of her husband's other wives roasted live over an open fire. (Alexander, it's said, was very mad at her for the roasting. But he got over it. He loved her.)

Whether Alexander was a born killer I couldn't tell you. But he seems to have shown he was his parents' child early on. Before he was into his teens he is said, by some accounts, to have murdered his astronomy tutor. Later, he murdered rivals to the throne he inherited from his father. By the time he himself died he is thought to have killed more people than anybody else in history ever had up to that time.

In one battle alone, says Plutarch, Alexander's army killed 110,000 Persians. Plutarch leaves the impression this was a considerable achievement.

Whether the Persians felt the same way he doesn't say.

Plutarch, incidentally, probably exaggerated the death toll. One expert estimates that in this battle Alexander probably killed only fifteen thousand Persians. In the old days writers tended to inflate the casualty figures.

Whether he enjoyed killing is unknown. But he seems to have had a pretty high tolerance of it. Supporters point out, though, that he always killed people in the open. Alexander was like that. There wasn't a sneaky bone in his body. If he wanted you dead, he came right out with it. Nobody he killed ever died wondering who'd done it.*

When he killed the wrong person, he was always very sorry. Plutarch says when Alexander killed his best friend during an argument he deeply mourned the loss, crying his heart out for two whole days.

Plutarch says Alexander slaughtered people to show them who was boss. His apologists, however, claim he was a good man all in all. Biographer Sir William Tarn explained that Alexander was driven in his conquests by the mission "to do something to outlaw war." Another scholar, W. A. Wright, has said of Alexander: "He boldly proclaimed the brotherhood of man."

---

* "Throughout his reign, Alexander never stands suspect of a surreptitious killing. When his power was vast, and he could have had anyone he chose put quietly out of the way, he suffered annoyance, frustration and downright insult from men he heartily disliked or distrusted; nothing happened to these people till he was ready to proceed against them openly." Quoting Mary Renault, *The Nature of Alexander* (1976), p. 66.

Did he cut the Gordian knot? Most people don't know what the Gordian knot was, but they know he cut it.* Scholarly opinion is divided on the matter. Some say he untied the knot. Others say he cut it with his sword. And some claim the whole story's nonsense, that there was no Gordian knot and that Alexander didn't untie it or cut it.

He finally stopped conquering people after nine years in the field. It came about one day as he was preparing to cross the Beas, a river in India. Alexander shouted, "Let's go." And his men shouted back, "Forget it." And that about ended it, as Alexander wasn't much of a conqueror without an army.

Why did his men refuse to go further? It may be they were simply homesick. Or they may have been tired of the rain. But biographer Peter Green is of the opinion that they'd finally figured out that Alexander's aim was to conquer the whole world. And they didn't want to.

Alexander died when he was thirty-two. It was probably just as well. With his army unexcited about new conquests, there just wasn't much to live for. Nobody, incidentally, knows how he died. He may have been poisoned. Or he may have partied too much. He died after a two-day feast.[3]

---

* It was a complicated knot that tied together the yoke and a wagon owned by Gordios, a peasant farmer who lived in Gordium. It was said that the man who untied the knot would become the lord of Asia. The wagon, incidentally, is said to have been used by Midas, the famous leader with the golden touch.

HERODOTUS

What of Herodotus, the Father of History?

Herodotus's method in writing his books was to include: (1) every story he ever heard, whether it was true or not (like the story about ants as big as foxes), (2) made-up Persian speeches, (3) plagiarized texts, and (4) out-and-out lies. And people decided to call him the Father of History. Makes perfect sense, doesn't it?[4]

CAESAR

First off, nobody ever called him Julius Caesar. They just called him Caesar. They didn't use first names back then unless a man had male siblings from whom he had to be distinguished.

That the cesarean section is named after Caesar is inaccurate. The name comes from the Latin word *caedere,* meaning "to cut."

Whether he himself was born through a cesarean section is probably nobody's business. Anyway, the experts can't seem to agree whether he was or he wasn't. They don't even agree on whether the operation was performed at the time.[5]

Almost everybody is aware that he was an emperor—or thinks he was. But he wasn't. Roman leaders weren't called emperors for another generation, even though Rome had already conquered half of Europe. In Caesar's day Rome was still formally a republic, though Caesar himself helped bring about the fall of the republic by inserting the army directly into politics.

So what was Caesar? He was dictator for life. His apologists like to point out that unlike modern dictators, who usually appoint themselves to the position, Caesar was appointed dictator by the senate, as provided for under existing Roman law. But while it was all just about legal, the senate would seem to have been influenced a little bit by the fact that he had the army behind him.

A lot is usually made of his decision to cross the Rubicon, as well it should. His decision ended, in effect, five hundred years of republican rule. Afterwards, the army dominated Rome. But it's interesting that for all the talk about the Rubicon, nobody knows where it was. All we know is that it was one of the streams marking the border between Italy and Gaul near the Adriatic coast.

Caesar is credited with the phrase "The die is cast," which he is supposed to have remarked as he prepared to cross the Rubicon. But Caesar did not coin the phrase. Plutarch says it was common even in Caesar's day.

Caesar deserves to be remembered as the bragging author of the line, "I came, I saw, I conquered," which he wrote in a letter quoted by Suetonius.

But when he died he did not say, "*Et tu,* Brute?" What he said—Shakespeare notwithstanding—was: "And thou, Brutus, my child!" (Caesar believed that Brutus was his son. He had an affair with Brutus's mother lasting some twenty years.)

That he died on the Ides of March is true. But the movies are wrong in suggesting he ignored the warning to stay away from the senate. Actually, upon being warned he immediately decided to postpone his appearance. But Brutus subsequently persuaded him to go.

What I find most interesting about Caesar is not what's said about him, but what's not said about him. Thus, scarcely anybody ever recalls that he, like Alexander, was one of the world's great killers. Pliny estimates that in Caesar's Gallic campaigns alone his army killed 1,192,000 people. Undoubtedly, this is an exaggeration. But nobody seems to question that he killed a lot of people, whatever the exact number.[6]

CLEOPATRA

First of all, she wasn't Egyptian. She was Greek. Her family had lived in Egypt for three hundred years or so, which might make her Egyptian in your eyes and mine, but to the Egyptians she was still Greek.

What she is famous for, of course, is her love life. But rumors of her promiscuity are unfounded. Lucy Hughes-Hallett, author of a highly esteemed book on the myths of Cleopatra, says Caesar and Antony were her only lovers. Some have claimed she was so great in bed men agreed (literally) to die for the opportunity to spend a single night with her, but this was just a lot of talk.

Blaise Pascal is the one who said that if her nose had been shorter the history of the world would have been different. But he was mistaken. She got her way with Caesar and Antony because of her wit and charm, not because of her looks. By any age's standards she was plain. She had an ungainly hooked nose and a fleshy face. You can see her face on the Roman coins Antony made in her honor. Elizabeth Taylor she wasn't.*

She *was* cunning. She actually arranged to get in to see Caesar by having herself rolled up inside an Oriental carpet presented to him as a gift.**

But she did not cast a spell over Caesar. He stayed on in Egypt because Egypt was rich and he needed the money ("the civil wars had been expensive"). He probably fell in love with her, but he never forgot why he was there. It was to get his hands on her fortune, which he claimed was rightfully his anyway because of a debt run up by her father. As debts go, it was rather a large one: 6,000 talents, an amount approximately equal to Egypt's entire annual revenue.

Caesar *was* romantically interested in Cleopatra—enough so to see that she got a proper marriage. To someone else. It gets even more complicated. The person he wanted her to marry was her brother. To paraphrase a line made popular by the humorist Will Cuppy, it made sense if you were an Egyptian.

Caesar did make her queen of Egypt, but he probably didn't do it for the love. Historians surmise he didn't have anybody else to appoint. Any

---

* And no, she did not wear "pale peach lipstick," and she did not wear petalled bathing caps. Nor did she wear bangs. See p. 264.

** Caesar never mentions the carpet story in any of his voluminous writings, leading some historians to doubt it. But it's not exactly the kind of story Caesar liked to tell. If a story didn't end with somebody's head hacked off, he wasn't much interested in it.

Nor, incidentally, did he ever acknowledge fathering Caesarion, the child she claimed was his.

Roman he gave the job to would instantly have become a potential rival. To a man like Caesar this didn't seem too attractive.

Her relationship with Antony was about the same as with Caesar. It could be very romantic. Like the time they had their celebrated meeting at Tarsus, in Asia Minor, when they are said to have fallen madly in love at first sight.* First, they made love. Then Antony agreed to kill her sister so Cleopatra wouldn't have to worry about any challenges to her authority. Then he went back to his wife. As I said, it was truly romantic.

No doubt Antony felt true passion for Cleopatra, just as Shakespeare says he did. He just had a hard time working things out so they could be together. As it happens, his wife died just about this time, but he declined to marry Cleopatra as he had decided it would be better if he married someone else, Octavia. Of course, as soon as he found the time he ran off to see Cleopatra and the twins he'd fathered. But Cleopatra was miffed. It had been three and a half years since he'd visited.

The critics said he'd been so smitten with her he'd do anything to be with her, even if it meant putting his career and reputation at risk. But every step of the way he seems to have put his career first. It wasn't just an accident that he'd married Octavia instead of Cleopatra. Marrying Octavia, his rival's sister, helped him maintain his political position. And it wasn't an accident that

---

* It wasn't actually love at first sight. They had known each other for years. But why spoil a good plot?

he finally took up with Cleopatra just at the precise moment when he needed her treasure and her navy.

Whether he was crazy in love with her or not, everybody agrees he hurt his reputation when he took up with her. Romans didn't like a Roman cavorting with a Greek queen and they made up a lot of stories about the two of them, like the one about the pearl, which was supposed to show she was wickedly decadent. The story is that one day at a banquet given in Antony's honor Cleopatra dissolved a monstrously expensive pearl in a cup of vinegar. But if it was true, as one wag has commented, then "vinegar was different in those days from the present-day kind." Pearls don't dissolve in vinegar.

As crazy as he was supposed to be about her, she was supposed to be even crazier about him. Her love is said to have been so strong she couldn't live without him. Thus, at the end of the movie, when he dies, she kills herself.

The romantics may not want to hear this, but the truth seems to be that she killed herself because she'd heard she was going to be paraded in disgrace through Rome in chains. Her grief, real as it undoubtedly was, had nothing to do with her decision. We now know that she was not going to be paraded through Rome, in chains or otherwise. She was just led to think she would be so that she'd commit suicide.

Plutarch is responsible for the story that she died from an asp's bite, but he didn't say it was true. It was just one of those stories he had picked up. All we know for sure is that she had two tiny marks on her arm when she was found dead. The business about the asp may have arisen simply because asps were a common emblem of Egyptian royalty.

Incidentally, she was supposed to have been a terrific queen: efficient, prudent, dedicated.[7]

CALIGULA

Although historians long ago gave up the practice of dividing leaders into the good-hearted and the hard-hearted, most people never have. They need their heroes and villains and they need them pure. Which brings us to: Caligula, "the Roman emperor people love to hate."

Much of the case made against him, however, rests on histories by Suetonius and Dio Cassius, and much of their information is manifestly unreliable. That he threw criminals to the lions, that he had incest with his three sisters, that he wanted to appoint his horse as a consul, all of this comes from Suetonius and Dio Cassius and all of it is unsubstantiated. It is especially unfair that the story about the horse has gotten the play it has, for even Suetonius, the source of the story, put it down as mere rumor.

Whether he was "mad, bad or ill," as the historians put it, is still in dispute. But the experts do not think he was all three, which is, one gathers, what most people think. As for whether he was certifiably mad, it's difficult to say nearly twenty centuries later, but the evidence is sufficiently ambiguous that the two major biographies published in the last fifty years conclude he wasn't. Dr. Anthony Barrett, his most recent biographer, says that while Caligula was certainly self-centered and arrogant, he was "capable of rational decisions, capable of statesmanlike acts." And if he seemed self-indulgent and egotistical, who, made ruler of the Roman Empire at age twenty-four, wouldn't be?[8]

NERO

Nero you can't go wrong hating. He was vulgar, cruel, salacious, greedy, self-important, and ignorant. In other words: your typical emperor, only a little worse maybe.

But bad as he was, he wasn't all bad. True, he killed his mother. But to be fair you must remember he did not kill his father. True, he killed his first wife. But most of the women he slept with, I believe, he let live. True, he did debauch one vestal virgin. But he never laid a finger on the other five.

Besides, he loved to sing and dance.

Of course, the big event of his reign was the week-long fire in Rome. This would have been a problem for most other emperors, but Nero had his fiddle and well, this seemed like a good time to play, seeing as how nobody would expect the emperor of Rome to grab a bucket and help out, right?

I know this sounds terribly plausible, but just because Nero was corrupt doesn't mean he was stupid. The fact is he did not fiddle while Rome burned. He didn't even own a fiddle. He owned a lyre.

Nor did he have anything to do with the fire starting. He was fifty miles away when it began.

Anyway, it so happens he behaved himself during the crisis. He opened shelters for the homeless, reduced the price of corn, and had food brought in from the provinces.

Not that he didn't make a mistake or two. So he started a huge rebuilding project that couldn't be finished and cost a fortune. So he blamed the Christians for the fire to get people to stop blaming him. So he persecuted a few hundred people. If you were Nero and you thought like he did, you'd have done the exact same things.

He wasn't universally unpopular. When he died, historians say, people even threw flowers on his grave. But, of course, people always throw flowers on the graves of dictators. There are a lot of stupid people in this world. Some Russians remain devoted to Stalin, some Germans to Hitler. So it would probably be a mistake to read too much into the story about the flowers. Most Romans, I'd say, were happy Nero was dead.[9]

THE FALL OF THE ROMAN EMPIRE

Rome fell, to be sure. It just didn't fall when it was supposed to. All the reference books say it fell in A.D. 476.* But Romans didn't know this, and kept the empire going another two centuries or so.

Why have we all been taught to believe it ended in A.D. 476? Because one day, about three hundred years ago, historians decided it would be easier for students if world history were divided into three periods: Ancient, Medieval, and Modern. And they figured that 476—the year of Rome's last emperor—was a nice date to use in marking the end of an epoch. But the selection of 476 was arbitrary. As historian Richard Haywood notes, Rome had been without an emperor before and done fine, and it did fine this time as well.[10]

Why did Rome fall? Was it because Christianity weakened the bonds that had held it together? Was

---

* *The Timetables of History* (1979), p. 30; *The Concise Columbia Encyclopedia* (1983), p. 728; etc., etc.

it because people became corrupt?* Was it because it just got too big? Was it because of the barbarian attacks? Was it because they had started using lead pots and got lead poisoning? (Yes, even this argument has been advanced.) Or was it simply that empires always fall and somebody decided this was as good a time as any?

The correct answer is, of course, that none of these answers is correct. There wasn't any one single cause.[11]

An underestimated factor may have been that they made too many stupid mistakes. Take Hadrian's Wall, built in England at the time of Emperor Hadrian. A prudent government, concerned with the defense of the wall, would have installed a moat around the outside. But what did the Romans do? They built moats on both sides of the wall, at a cost, it is said, of a million days' labor.

Why did they build the inside moat? Historians have put forward a lot of fancy explanations, one being that an inside moat was a convenience for the customs officials. But the chief conclusion, I think, is that the Romans did it because of stupidity, a conclusion they themselves seem to have reached a short time later when they decided to fill in the inside moat.[12]

---

* American politicians like to make the argument that because Rome was corrupt and fell the United States could well fall too because it's fast becoming corrupt. Me, I don't know what's going to happen to the United States. But I know this: Rome did not fall because of corruption. The Roman Empire, except for one relatively brief period, was always more or less corrupt.

That the collapse of the Roman Empire was a calamity is true. Seeing all the bad that came of it—the sacking of Rome, the destruction of art, the withering of great cities, the deterioration of the system of roads, the ruin of Mediterranean trade, and the loss of European unity—it's difficult to imagine any good that came of it. But some good did result. The break-up of the empire led to the abolition of slavery in Europe. Of course, this, in turn, led to the birth of serfdom. But the slaves were better off as serfs than as slaves.[13]

Incidentally, did you ever wonder why historians always refer to the sacking of Rome as "the sacking of Rome"? Nobody says Watts was sacked or Los Angeles was sacked, but Rome, it was sacked.

Who sacked it? Everybody always thinks it was the barbarians alone who sacked Rome. But they got a lot of help from the slaves. In fact, the slaves probably did more damage to Rome than the barbarians did.

The sacking of Rome, in any case, is overrated. It wasn't the catastrophic event it's been made out to be. You know when ancient Rome was really destroyed? It was during a wild building boom in the Renaissance.

It was like Vietnam. To save the place they had to destroy it. Take St. Peter's basilica. This great edifice, "the oldest, largest, most sacred building in Christendom," survived for 1,200 years. Then the Renaissance came along and it was leveled.

Why? Americans will be delighted to hear this: it was because the Romans wanted something new. They were so proud of old Rome they wanted to hurry as fast as they could and rebuild it.

Whole sections of the city were demolished, sections that had survived the barbarian raids, the revolts of the slaves, numerous wars, and all manner of other calamities. It came about this way. Say you were building a brand new courthouse and you wanted to put in a couple of columns, nice marble ones like the kind they used back in the good old days. Where would you go to get them? Why, you'd take them from some old building somewhere.

Michelangelo and some others complained about the practice, but nobody listened.

What they did with Rome's old statues, incidentally, is even more appalling. They used them to make lime! Ever wonder what happened to all the thousands upon thousands of marble statues made in ancient times? In the Renaissance they burned loads of them to make lime. Why lime? Well, they needed the lime to make plaster. They could have quarried new marble to make plaster, of course. But this was easier.

You mustn't think it was just the Renaissance Romans who burned statuary to make plaster, though. Romans continued to do it later as well. One horrified archaeologist, in 1883, reported seeing old Roman statuary being burned to make plaster in a kiln near the Atrium of Vesta. Eight of the statues, he observed, were "nearly perfect."[14]

THE BARBARIANS

The difference between a Roman and a barbarian was what, really? They both thought a fun afternoon was going to the Colisseum to watch defenseless animals get slaughtered. They both enjoyed seeing gladiators hack each other to death. And they both thought life was so monotonous a little war now and then was a good thing.

Religion? The Romans worshiped the sun god. The barbarians worshiped things like trees. Of course, later the Romans adopted Christianity. But, then, so did the barbarians. Admittedly, they did not share the same belief in Christianity. Romans believed Christ was "of the same divine nature as God." Barbarians, like many early Christians, did not. You can see how this made a really big difference.

They both believed in persecuting people with whom they disagreed, but there *was* a difference in the way they went about it. The Romans believed you should only persecute someone after you'd given them a trial. The barbarians believed you didn't have to bother with a trial.

Personal hygiene? Here a real difference existed.

Romans believed in taking baths, barbarians, on the whole, did not.

Oh, and the barbarians preferred living in rural villages rather than in cities.

Of course, I am speaking here as if all barbarians were alike. This wasn't so. There were your average barbarians—your Goths, your Visigoths, your Franks, your Vandals—and then there were the Huns. The Huns were *bad.* Everybody hated them.

Did the barbarians—I mean your average non-Hun barbarians—want to see the Roman Empire destroyed? No. The reason they were always invading the empire, places like Gaul and Spain, was to get away from the Huns.

Take the time the Visigoths flooded across the Danube. They wanted protection from the Huns, who were bearing down on them from the north. In exchange for land, they even agreed to help defend the empire from the Huns. Indeed, the Roman army in time came to be made up mainly of barbarian forces. A barbarian was even put in charge of the army.

Not to say there wasn't trouble between the Visigoths and the Romans. But it wasn't the Visigoths' fault. The Romans took advantage of them, like selling them food at ridiculous prices. This made the Visigoths a little mad.

Why do we think the barbarians were BARBARIC (I mean savage)?

It wasn't because of anything the Romans said. The Romans, more or less, held the barbarians in high regard.* In Roman times the word barbarian didn't even have a negative connotation. Anybody who wasn't Roman was a barbarian, even if they had a Ph.D.

The awful truth is we were misled by the Renaissance humanists. They were good at painting but not so hot at history. (It was they who first turned the name of the Vandals into a synonym for crime.)

Incidentally, do you know why we call Gothic architecture Gothic? It's because the Renaissance humanists thought it looked ugly and the worst name they could think to give it was Gothic, the Goths being barbarians and all.[15]

---

* Tacitus: "[The barbarians are] purer and more chaste [than we Romans]."

PART 2

THE DARK AGES

IGNORANCE

THE CRUSADES

KNIGHTS

HUNDRED YEARS' WAR

SHYLOCK

THE SPANISH INQUISITION

Knights

IGNORANCE

Fulfilling students' worst fears, the Fall of Rome was immediately followed by the Dark Ages, which nobody yet has ever found the least bit interesting. But be that as it may, it is part of history and must be looked into.*

The Dark Ages are said to have begun one morning, not long after the Fall of Rome, when everybody suddenly woke up dumb. And as people were dumb they forgot how to turn on the lights, and therefore lived in darkness for a thousand years.

Or so we were led to believe. Actually, it now appears people didn't live in darkness for a whole thousand years at all. At worst, we are told, the period of darkness lasted just five hundred years. And many historians insist it's misleading to think of any part of the period as dark. Which is why historians prefer nowadays to refer to the era between the Fall of Rome and the Renaissance as the Middle Ages.[1]

---

* As a favor I promise not to mention feudalism, desmesnes, or subinfeudation, especially as I am confused about these items myself.

Unfortunately, people are so thoroughly brainwashed they cannot stop thinking of the Middle Ages as the Dark Ages. Test yourself. Try using the term "Middle Ages" without thinking to yourself "Dark Ages." Impossible, isn't it?

Shockingly, the brainwashing continues to this very day in history classes all across this great country. The reason? Historians insist it's important to indoctrinate students in mistaken interpretations so they'll appreciate the new interpretations. This is known as teaching history by the Confusing Method.* Which leads to lesson plans like this:

MONDAY: Tell students all the old malarkey about the Dark Ages. Include the stupid stuff: people were superstitious, practiced witchcraft, argued about the number of angels who can dance on the head of a pin, etc., etc., etc.

TUESDAY: Tell students why everything you told them on Monday was wrong or misleading.

What, then, were the Middle Ages really like? Unfortunately, nobody has yet figured this out. Which is why in this book we will follow the time-tested method of saying what the Middle Ages were not like.

To begin with, they were not uniquely superstitious. Superstitiousness was as bad under the Roman Empire. From whom do you think medieval

---

* Humorist Dave Barry calls it the Boring Method. He may have a point.

people learned all those old superstitions, anyway? From the Catholic Church? The church opposed superstitiousness as a remnant of paganism. Protestants subsequently criticized church leaders for not doing more to root out superstitiousness, but Protestants weren't too successful in eliminating it either.[2]

Witchcraft? To be sure, people in the Middle Ages believed in witchcraft. But they didn't go around burning witches. That came later, after the Middle Ages ended, and the authority of the Catholic Church had eroded. During the Middle Ages, if you thought someone was practicing witchcraft you simply turned the offending sinner over to your local Catholic clergyman. What the clergyman said to the alleged witch I don't know, but I think it was something like, "You better stop playing with witchcraft, OR ELSE!" And that was usually enough, as even a witch could take a hint.[3]

That philosophers wasted their time debating the number of angels that can dance on the head of a pin is untrue. What they wasted their time debating was whether angels defecate (and other stupid questions).[4]

That classical learning died out is sort of true. It died out (for a time) in the old Western Roman Empire but not in the Eastern Roman Empire, where civilization had continued to flourish, nourished by stimulating contacts with the Arabs. I should explain what the Eastern Roman Empire was. It was the part of the Roman Empire centered in Constantinople that did not fall for another thousand years, but that you never hear anything about for some reason.

That Catholic monks helped keep classical knowledge alive in the Middle Ages by assiduously copying the ancient texts is flat untrue. In fact, the church engaged in a systematic campaign to suppress the classics, being that the classics had been produced by pagans. Many monks weren't even literate. Of the thousands of monks who lived in the thirteenth century at the Swiss Abbey of St. Gall, for instance, not one was able to read or write.[5]

Another thing you never hear anything about is the twelfth-century renaissance, which occurred right in the middle of the Dark Ages.* It was at this time that Gothic architecture was invented, Oxford was founded, Aristotle was rediscovered, and experimental science was inaugurated. So what happened? Unfortunately, it wasn't yet time for the real Renaissance, so after two centuries of fun they had the Black Plague and everybody went back to being glum.

We do hear about Charlemagne, the great military leader (claimed by both the French and the Germans), who conquered the Saxons, the Lombards, and the Bavarians in the late eighth and early ninth centuries, and who came to the rescue of the pope. But while Charlemagne is remembered, his legacy, the Carolingian empire, is hardly known, though out of it came the Holy Roman Empire.

Of late it's been argued (by Kirkpatrick Sale, the left-wing author of a book attacking Columbus) that the popularity of bullfights, cockfighting, and

---

* Which is why you never hear anything about it. To have a renaissance come anywhere but at the end of a dark age is confusing, so teachers just skip over this part.

bear-baiting in the Dark Ages is proof of their darkness. For it was, he says, in its treatment of animals "that the medieval world truly revealed itself." He neglects to mention that the Romans treated animals even worse, which presumably would make their period even darker.[6]

I come now to the question of the economy in the Middle Ages, as the subject can't be avoided even if it is deadly dull. The biggest development was the invention in the tenth century of the rigid horse collar. This was to the tenth century what the invention of the automobile was to the twentieth. I know this is hard to believe but it's what the experts say, so argue with them if you want. Anyway, it's supposed to have revolutionized agriculture. Also at this time people discovered that food tastes better with spices, so they started trading again. And in no time at all, say historians, living standards began to rise dramatically (until, of course, the Plague hit).[7]

Why, if the Dark Ages were not really all that dark, did anyone ever think they were? Once again the fault lies with the Renaissance humanists. They got this one wrong, too.[8]

THE CRUSADES

Whether the Crusades should be thought of as: (a) a noble adventure, or (b) "the most signal and durable monument of human folly that has yet appeared in any age or nation,"* is a matter of opinion.

Me, I'm not so crazy about them. Take the First Crusade. This is the one started by that hero of the Holy Wars, Peter the Hermit. The highlights? First, his people decided it would help if they killed a couple of thousand Hungarians. Then, for the hell of it, they went after Germans. Then they went after Greeks. And along the way, they got in the killing of some Jews: in Speyer, killing 12, in Worms, 500, in Mainz, 1,000. Invading Jerusalem gave them the opportunity to kill some more: mainly Moslems but Jews, too. *And this was the crusade that's said to have been a success.*

On the Second Crusade they pillaged Byzantium.

On the Third (this was the one with Richard Lion Heart), they massacred three thousand in-

---

* David Hume.

nocent Moslem villagers (including women and children).

On the Fourth they sacked Constantinople.

On the Fifth they were caught in the rising flood waters of the Nile and forced to flee.

Between the Fourth and the Fifth Crusades came the Children's Crusade. It was made up of two contingents: 30,000 children from France and 20,000 children from Germany. The French children traveled by ship from Marseilles to Alexandria, and were promptly sold into slavery. The German children marched across the Alps, got homesick, and deserted. Many died.

And so on and so on.

In the end, what came of it all? Well, the Christians finally got tired and went home and the Holy Lands reverted to Moslem control.

Why had so many agreed to join the Crusades, of which there were nine in all? Everybody had their own reasons, and some of them were probably pretty good reasons, too, or seemed so at the time, but who's to say, really? The problem with human beings is you can never say for sure why they do stupid things. They just do, that's all.

Knights probably had better reason than most to join. Going on a crusade was a good career move. Hop on a horse, kill a few hundred people, and come home a hero. Of course, there was always the chance you'd end up dead, but people thought differently then. So you died, so what?

In 1366, when 800 more knights than were needed showed up at a recruiting office and were sent home they got in a foul mood. Why? Because, Lord d'Albret explained, "they were all set and ready to go abroad to Prussia, to Constantinople, or to Jerusalem *as every knight and squire who wishes to advance himself does.*"*

Besides it being a good career move, joining a crusade was also good for the pocketbook. And as many knights were broke, this seemed like a good reason for going. (Why were so many knights penniless? Because in medieval times it was the eldest son who inherited his family's wealth. This left a lot of younger sons in the poorhouse.)

What was it like to go on a crusade? I couldn't say precisely, but the crusaders stole their food from local farmers, ran around in mobs, and had sex (the men are said to have brought along thousands of mistresses and prostitutes).[9]

---

* My italics. (They didn't know from italics in 1366.)

KNIGHTS

Was the knight's life romantic? Well, they lived in cold, drafty castles. In the field they had to put up with "heat, cold, fasting, hard work, little sleep and long watches."* They often died young. They rarely rescued damsels in distress. And many died broke. Says one historian, they spent more of their time "in search of income than romance."

They got to play in tournaments, though, didn't they? Most didn't. Tournaments did not become popular until the late Middle Ages.

Anyway, tournaments weren't as romantic as Hollywood has made them out to be. Often there were "accidents." Like the time knight Roger de Lembum "accidentally" forgot to use a blunted lance instead of a sharpened one and killed his jousting opponent. This kind of thing happened so often that eventually both the Catholic Church and the French monarchy supported a ban on tournaments.

They wore suits of shining armor, didn't they? Actually, through most of the Middle Ages knights

---

* Geoffroi de Charny, in 1352.

wore plain suits of wire mesh. Suits of shining armor weren't developed until near the end of the period. (And anyway, most knights couldn't have afforded them.)

They did live by the code of chivalry. This part is true. But the code could be peculiar. Say you were a knight and you wanted to kidnap someone with whom you were at odds. Under the code, you could. All the code said was that your demand for ransom had to be "reasonable." What was a reasonable ransom? It was whatever you decided it was. That was the beauty of the code. It was flexible.

Another peculiar feature of the code was that gentlemen were required to behave gentlemanly only toward other gentlemen. If a knight wanted to give a peasant a good thrashing he could.

Another myth about knights is that they always fought on horseback. English knights, at least, often didn't. The French thought this was very stupid, and during the Hundred Years' War always had a good laugh every time they saw that the English were fighting on foot again. Ha, ha, ha, they'd go. "Look at those dumb Englishmen, fighting on foot." And then the English would win—always such a surprise for the French.

Why didn't the English fight on horseback? Because they often couldn't afford the horses. Besides, they discovered that archers armed with long bows could defeat horseback-riding knights. Henry V beat the French at Agincourt this way.

Most English knights did not fight either on foot or horseback. The fact is, most English knights never fought. In the words of historian Sidney

Painter, the "majority of English knights never got nearer to real fighting than paying their scutage."* It wasn't England's knights who usually fought and won England's wars in the Middle Ages; it was mercenaries. The advantage to the king in hiring mercenaries was that they could be hired directly in Europe, where most of the battles took place. Transporting English knights to Europe would have been expensive.

Besides, when the mercenaries died the king didn't have to worry about the reaction of their families. A king could lose a lot of mercenaries in stupid wars before the people back home began wondering whether they ought to get themselves a new king. When it was their own kind whom the king got killed, they weren't as lenient.

What became of knighthood after the Middle Ages? Englishmen continued to be named knights well into the seventeenth century, but by then it wasn't the honor it formerly was. One of the great complaints of Englishmen then was that they were being named knights and they didn't want to be. People became so afraid they might be the next one named a knight they forced Parliament to pass a law saying nobody could be made a knight who didn't want to become one.

The problem wasn't that they'd be made to fight and they didn't want to. As we've seen, most English knights never had to fight anyway. The problem was taxes. Knights in the seventeenth century paid higher taxes than other folks.

---

* Scutage was the fee levied by the crown on knights not employed in battle.

Forcing knights to pay higher taxes was one of the bright innovations of King Charles I. (It was dreamed up by his chief accountant, Julius Caesar.) At first, as you can imagine, it seemed to Charles like a terrific idea. Anytime he needed to raise more revenue he just named some more knights. But I'm sure even he eventually saw that it was a mistake, like around the time he was beheaded.

Now, I'm not saying they chopped off his head just because he'd introduced the special knight's tax. This was just one of many things people didn't like about his reign. But it didn't help.[10]

HUNDRED YEARS' WAR

Naturally, it didn't last a hundred years. Officially, it lasted a hundred and sixteen years (1337–1453), but a hundred must have sounded better. Who could remember the Hundred and Sixteen Year War? Besides, it's not as if they were fighting all one hundred and sixteen years anyway. They stopped fighting years before they officially agreed to admit that they'd stopped fighting. They waited before signing the treaty, however, because it's easier to sign a treaty to end a war if you wait so long no one can remember why it began.

Actually, this was the second peace treaty. In 1360 they'd signed another one. It was called the Treaty of Brétigny. You may not have heard of it, but at the time it was a big deal. Everybody had high hopes and celebrated. Naturally, war broke out again almost instantly.

The French, who had started the war again, had every intention of keeping the peace, I'm sure. They just couldn't help themselves. They had given up a lot: Gascony, Aquitaine, Calais. Naturally, they had second thoughts.

SHYLOCK

That most moneylenders in the Middle Ages were Jewish is widely believed. I suppose this is what you heard, too. So you think I'm going to tell you different? Nope. They were. Which is why to this very day Jews are associated with moneylending.

How did Jews come to dominate the field? It happened this way. Day after day the Catholic Church condemned moneylending, saying you shouldn't do it and if you did nobody decent would ever want to speak to you anymore and you'd rot in Hell. And Christians, believe it or not, got the hint. Thus did Jews obtain their "nefarious" grip on the world's financial institutions.

How long did their monopoly last? A couple of centuries, then the Medicis and some others got interested in banking and before you knew it, Christians were taking over the whole thing.

Why did Christians decide it was now okay to get into banking? It happened like this. One day some theologians looked up from their Bibles long enough to notice that credit seemed to be essential to the growth of the economy. And all of a sudden banking didn't seem so immoral anymore.*

That Jewish moneylenders were mean and nasty is doubtful. As historian Joseph Shatzmiller points out, to stay in business they had to smile to keep their customers happy. Then as now, banking was a competitive business. No moneylender who went around with a dour look on his face, or worse yet, looked like some kind of beady-eyed monster, would last long. Who'd want to owe money to a person like that? It's just a stereotype, and a mean one, too.[11]

---

* They still didn't approve of high interest rates. But who does?

THE SPANISH INQUISITION

The Spanish Inquisition *was* awfully bad, but history's full of horrors. Why pick on the Inquisition?

To be sure, they killed a lot of people—twenty-five thousand or so. But that was over a period of three and a half centuries.

You know how many witches were put to death in Europe over the same period? *Several hundred thousand.*

So the Inquisition was bad, but it could have been worse.*

They did torture a lot of people, but not as many as you might think. Take Valencia. Of the 2,354 people arrested there between 1480 and 1530 they tortured only twelve. Was Valencia representative? I don't know and neither does anybody else.**

---

* As arguments go, this isn't much of one, I know. But I thought it was intriguing and worth mentioning.

** I did see one report that about a third of the people arrested for crimes punishable by torture were tortured. The catch is, we don't know how many people were arrested for crimes punishable by torture.

It's true that the Inquisition mercilessly persecuted the Marranos—Spanish Jews who converted to Christianity but who continued to practice Judaism in secret—but this was mainly in the beginning. Through most of its history the institution was used by Catholics to hound other Catholics. Not that this is exculpatory, but I suppose it's something that they could go after their own kind with the same vehemence they went after others.

(Of course, the reason the Inquisition did not go after more Jews was because there weren't too many Jews in Spain to go after. The Spanish government had expelled them; only Jews who converted—or who pretended to convert—were allowed to remain. The expulsion of the Jews occurred in 1492, the same year Columbus sailed on his maiden voyage to the New World.)

And the fact must be faced that most of the people tried and convicted by the Inquisition probably were guilty of the crime with which they were charged. In most cases, to be sure, the "crime" was heresy. But back then heresy was considered by many to be a serious offense.

If the Inquisition was not one of the world's worst horrors, why does everybody think it was? It was because the Protestants wrote the history books.

Anyway, the Protestants themselves weren't slouches at persecution. There they were, going on about the Spanish Inquisition, when all the while they were busy killing witches: thirty thousand in England alone in the sixteenth and seventeenth centuries. And you were wondering why so many people were eager to leave England to go to America? Of course, they killed witches in America, too. But in America they only killed bad witches. Or so I heard.

Finally, it is a damn lie that the Spanish government made money off the Inquisition. Sure, they took in a lot of dough. If you work at it, you can earn a fair penny confiscating the property of heretics. But Inquisitions don't come cheap. The money taken in barely covered the cost of the operation. Costs eventually went so high, in fact, that they had to cancel public executions. The community feasts held afterward were breaking the budget.[12]

PART 3

A NEW DAY DAWNS

THE SCIENTIFIC REVOLUTION

COPERNICUS

GALILEO

SCIENTISTS ARE HUMAN

Newton

THE SCIENTIFIC REVOLUTION

The Dark Ages, we all know, were followed by the Renaissance, but we're skipping over that part. Which brings us to. The Scientific Revolution.

What of it? It never happened. An early twentieth-century textbook writer made up the whole thing. Nobody alive during the sixteenth or seventeenth centuries ever even heard of it.

So you're thinking: maybe it happened but people just missed it? Maybe, but what kind of a revolution was it if people could miss it two centuries running?[1]

Science, of course, was changing and changing in important ways at this time. But it was still mired in medieval practices and superstitions. Take the sixteenth-century cure for a bad kidney. It goes like this: Take three jugs. Fill them with the patient's urine. Bury the jugs underground. And lay a tile of some kind over the jugs so no dirt would get inside when you filled in the hole. (Everybody knew if you got any dirt in the jugs the cure wouldn't take.)

Sound scientific to you?

Or how about this? When the poet Thomas Flatman got a knife wound his doctor, in accordance with practices advanced by the Royal Society, the most august scientific body in England, *put the medicine on the knife.* Flatman himself the doctor didn't touch.[2]

What of the gurus of modern science?

Johannes Kepler, the founder of modern astronomy, moonlighted as the official court astrologist for the Holy Roman Empire.

Leibniz, the philosopher and mathematician, practiced alchemy.

And Newton? He didn't exactly believe you could turn base metal into gold, but he held the opinion that gold could be turned into other substances if it could be made to "ferment."

His apologists claim he didn't really believe in alchemy, he just kind of toyed with it for fun. Serious toying, it seems: his alchemical notebooks ran to more than a million words.

Newton, of course, remains an important figure in the history of science. Even if he did get the idea of gravity in an unusually quaint way, watching an apple fall out of a tree. But he wasn't much good at history. His chronology of world history, which he worked on for years, was based on the date Jason and the Argonauts sailed to find the Golden Fleece. Nobody ever had the heart to tell him Jason was a figment of Greek mythology.

The story of Newton and the apple *is* true, incidentally, silly as it sounds. For a long time, it's worth noting, skeptics disbelieved the apple story. It had been traced to Voltaire, who said he'd gotten it from his niece, who'd supposedly gotten it from . . . well, you get the idea. But in 1936 it was finally confirmed independently. That year a publisher brought out the memoirs of one of Newton's closest friends, W. Stukeley. And there, on page 19, was the old apple anecdote.[3]

COPERNICUS

Copernicus was an awfully important person, we are told, and I don't doubt it. For he started a revolution (the Copernican Revolution). But it was an odd sort of revolution. Kind of went by and nobody noticed.

Even Copernicus himself missed it. He was under the impression he'd be remembered for his theory of circularity, the theory that the planets go around in near-perfect circles. He would have, too, but for one minor drawback: planets don't go around in near-perfect circles.

What he *is* remembered for, of course, is his discovery that the earth revolves around the sun. But neither he nor anybody else at the time thought this was much of a big deal.

As you may be aware, a few years later the Catholic Church got very upset when Galileo made the same point as Copernicus about the earth and the sun. But nobody ever bothered Copernicus. Maybe because he'd had the good sense to delay publication of his theory until just before he died.

When his book finally did come out it was ignored. Astronomers would have liked it, I think, but there weren't any yet. This was a problem. Not until the next century did the astronomer Johannes Kepler discover Copernicus.

Some have suggested that Copernicus isn't the big cheese he's made out to be because he wasn't really the first person to suggest the earth revolved around the sun. In ancient times there was a fellow named Nicetas who said the very same thing. But Copernicus *was* the first one to prove it. This, I believe, is important.

Did I mention Copernicus was Polish? Nobody ever does. Newton we remember was English. Galileo we remember was Italian. But that Copernicus was Polish is some kind of big secret. Except in Poland.[4]

GALILEO

Galileo is the famous scientist about whom a lot of anecdotes are told, some of which are even true.

Galileo, for instance, actually dropped some weights off the Leaning Tower of Pisa, or anyway, off of some tower somewhere, to prove that falling objects of different weights fall at the same rate of speed. We know this because he said so, right in one of his notebooks. Dropped weights "off a tower," he wrote.[5]

I'm afraid we have to give up the story about the swinging chandelier, though. It didn't happen. He did not get the idea of the "isochronism of the pendulum" after watching a swinging chandelier at church one day. He just got it, that's all.

Now what the hell is the "isochronism of the pendulum"? Trust me. It wouldn't interest you.[6]

Which brings me to one of the strangest anecdotes ever told about a supposedly sane human being. I am referring, of course, to the anecdote told about Galileo's final day before the Inquisition. As Galileo was about to leave the courtroom, where he had just been forced to renounce the view that the earth revolves around the sun, he supposedly saw the chandelier jolt and remarked, "*Eppur si muove*" (nevertheless, it does move).

True or not true? Not true. But what interests me is why anybody would make up such a story. Who but a fool would take that somber moment to utter such a flip remark? Yet this is the story that is told about Galileo over and over and over again as if it redounds to his credit.

In 1992, incidentally, Pope John Paul II announced it was a mistake for the church to have put Galileo on trial. It's just my opinion, but I am inclined to believe the rest of us had already figured that out.

QUESTION: Since Galileo renounced his view that the earth revolves around the sun, why is he always described as a martyr to scientific truth? He caved in and recanted.

He didn't even have such a hard time of it during the trial. He stayed with his good friend the pope. He even had a servant. His sentence: to spend the rest of his life on his country estate in Florence.

Sure, if he hadn't recanted he could have been tortured or even burned at the stake. But he recanted.[7]

In passing, it is worth noting that it wasn't just Catholics who felt threatened by Galileo's pronouncements. Martin Luther criticized Galileo as a "madman" who, in his yearn "for a reputation," "would subvert the whole science of astronomy." "Scripture tells us," wrote Luther, "that Joshua bade the sun, and not the earth, to stand still."

See. The Catholics got a bum rap.[8]

SCIENTISTS ARE HUMAN

Scientists, it turns out, are human. They neglect to mention this in normal history books, but the fact is they are. Which brings me to the subject of this section: fraud.

Leibniz, Newton, Kepler, Mendel: all have been accused of fraud, and may very well have committed it.

The accusation against Leibniz, made while he was still alive, was plagiarism. I don't know if he was guilty of it or not, but the Royal Society of London decided he was and condemned him for it.

The charge against Newton, "deliberate fraud," was made in 1973 by scholar Richard Westfall, in a detailed report on Newton's use of numbers in the *Principia,* which is commonly described as the first scientific work in which numbers were supposed to have been handled with expert precision. Westfall discovered that Newton boldly "fudged" the numbers to match desired results.

It was kind of a game with Newton. Anytime a critic noticed that one number or another seemed out of sync with the rest he'd announce to his assistant that it was time again to play with the numbers. And off they'd go to find a new number that seemed to fit better. Newton called this the "mend the numbers" game.

Kepler's offense, boldly captured in a newspaper headline, "Numbers That Lied," was discovered by William Donahue and reported in 1990 in the *Journal of the History of Astronomy.* Kepler had always claimed that his theory about the elliptical orbit of the planets rested on mathematical calculations. Donahue demonstrated that the numbers had been invented to justify the theory. Kepler's apologists insisted the manipulations would not affect his reputation, and Donahue agreed they wouldn't and shouldn't: "So he fudged a little. That doesn't take him down a notch."

Gregor Mendel, the Austrian biologist known for his pioneering work in genetics, reported genetic ratios that he allegedly could not have seen in his plants and that could not have resulted "from accidents of sampling," according to an article published in the *Annals of Science* in 1936. No one knows, however, if it was Mendel who made the mistake or one of his assistants.

Louis Pasteur, the French scientist who discovered how heat kills germs, lied about his methods, "massaged" his scientific data, and stole an idea from a competitor. The deceptions were discovered in the early 1990s by Princeton historian Gerald Geison, who had the opportunity to check Pasteur's notebooks against his public statements. In 1881, for instance, Pasteur publicly claimed to have saved a herd of sheep from a deadly anthrax virus by developing a vaccine that used oxygen to weaken the organism. In actuality, he weakened the anthrax with chemicals, an approach he "borrowed" from another scientist. In Geison's opinion, Pasteur "deliberately deceived the public" in an incident that constituted a "clear case of scientific misconduct." (But to give Pasteur his due, eventually an oxygen-based vaccine was developed that was superior to the chemical one. So Pasteur was on to something.)[9]

PART 4

THE FACTS OF LIFE

SEX: I

SEX: II

SEX: III

Casanova

SEX: I

So much is made of the importance of "sexual morality" to the survival of western civilization that one might almost believe that people in the West once were "sexually moral." But there's not much proof of it. Everywhere historians have looked they've found evidence of lustiness. Take western Europe, for instance. Historians report that they have been able to find only two brief periods in the last five hundred years when people in western Europe may be said to have behaved, by and large, "morally." These were during the high tides of Puritanism and Victorianism. And even then people did not march in moral lockstep. If they had the Puritans and the Victorians would hardly have found it necessary to denounce licentiousness as often and as vehemently as they did.[1]

Whether Elizabethans had sex as often as people say, I don't know. They didn't take sex surveys then. But they talked (or at least wrote) a lot about sex.

They also got into a lot of trouble over sex. Elizabethan church records in the English county of Essex, for instance, reveal that between 1558 and

1603 one in four adult church members was accused of a sexual offense. Historian Lawrence Stone estimates that about half of the accused were guilty. Offenses included fornication, adultery, incest, bestiality, and bigamy.

Promiscuity was so common in the 1500s in England, Wales, and Ireland, according to historians, that people didn't even feel ashamed of it. Professor Christopher Hill reports that "illegitimacy carried no social stigma."

Actually, people then were not less moral than they were at any other time. But it was the custom in the 1500s to marry rather late. This left people single a lot longer than was good for them.

I've always thought Elizabeth was the wrong queen for the times. That it was Elizabethans who got the Virgin Queen strikes me as a little odd. I think they would have been much happier under her father. Henry VIII would have understood them better.

A curious fact about the institution of marriage in Britain is that it doesn't have nearly the tradition behind it that people think it does. It wasn't until the twelfth century that the English began getting married in church. And it wasn't until the eighteenth century that they were required to by law.

The fact is that through most of English history nobody much worried if a couple was formally married or not. Or I should say among the lower classes nobody much worried. Lacking property, the members of the lower classes did not need to sanctify cohabitation with fine legalities. The rich, of course, did. For them marriage constituted an exchange of property.

How many people actually got married in church in the Elizabethan era? You will be shocked by the answer I read in one book. Only fifty percent.

The problem with church weddings was that they were hard to undo if the couple found out later they'd made a mistake. Church weddings also cost a lot.

The only alternative to having a church wedding was to have what was called a "private wedding." Why they called them this I don't know. They were anything but private, seeing as how they were usually held in a public tavern.

Private weddings were especially common among the coal miners in South Wales. Coal miners moved around a lot, and they often preferred to leave their wives when they did. Nobody thought this was the least bit strange.

It's said that when a coal miner abandoned his wife she would simply go to the beauty parlor, get a make-over, and advertise that she was available again. And in no time at all another man would take her up. Nobody worried that she'd been deflowered. A woman who'd been abandoned was, wrote a Welshman, "no worse look'd upon among the miners than if she had been an unspotted virgin."

Historians, incidentally, were dismayed to discover that lots of people got married in private. It meant they had to throw out all the old generalizations made about English mores. The generalizations had been based on the unrepresentative church wedding records.[2]

SEX: II

Having had about all the fun a person can, Englishmen settled down after the Elizabethan era and kind of went to sleep. When they awoke it was time for the Age of Reason, so named because people decided to live life reasonably. This meant giving up hang-ups about sex.*

The Age of Reason was actually a misleading name for the period, as it's led generations of students to think all people did back then was think big thoughts. It would be more honest to call it the Age of Sex.

The fact is too much attention is usually paid to guys like Voltaire, Rousseau, and their ilk, and too little to sex fiends like:

Augustus the Strong, the king of Poland, who is

---

* I'm skipping over, you may have noticed, the English Civil War. Even the English don't spend too much time on it. It's almost like it never happened. (Quick: How many people died in the English Civil War? See. Nobody knows anything about that damn war. The only person I ever heard died in it was Charles I. What kind of war is that?)

known to have fathered more than 350 children.

Frederick the Great, who is said to have engaged in all-male sex orgies.

And Gian Gastone, the last Medici prince, who is said to have kept a stable of 400 male and female sex slaves.*[3]

Why is it teachers don't usually mention the sex fiends? I think it's because they're afraid of the questions students might start asking. Like:

Is it true Catherine the Great died while having sex with a horse?

Was de Sade as bad as they say?

Was Casanova for real?

Was Byron writing from experience or did he just make up all that love stuff?

I can see where it might be troublesome to answer questions like these, but personally I have no problem with them. So:

Catherine the Great? That she died while having sex with a horse is a lie. Nobody's proven she ever had sex with a horse even once. She *is* known to have had sex with at least ten men.

De Sade? He wrote about killing people but his biographers now assure us he himself never did. He whipped quite a few and made them bleed badly enough to be hospitalized, but he never killed anybody.

As a husband he left a lot to be desired. Five months after his wedding he was arrested for whip-

---

* Teachers should bring up Gastone. If it got around that there were guys like him in the history books, people might read more history books.

ping a prostitute and masturbating on a crucifix.*[4]

Casanova? Sure he was for real. But contrary to popular belief, he didn't just sleep with women. He also slept with men. As long as he slept with someone, that's all he cared about.[5]

Byron, too, slept with men, but he never wrote about it. Three love affairs with men have been proven beyond doubt, but there may have been even more. It's been suggested one reason he came to have a "hectic involvement with women" was because he'd been with men and felt guilty.**[6]

The Age of Reason was also an important era for pornography. It was in the eighteenth century that the masses were first exposed to the stuff. Nobody ever mentions it, but one of the first things people began reading once they learned how was pornography.

A related development was the invention of the newspaper sex advertisement. One advertisement in a London paper enticed couples to reactivate a dull sex life by renting a "celestial bed" in the "Temple of Hymen." Cost: fifty pounds a night.

The English may very well have been the first people on earth to advertise sex in the papers. But I've never once heard them take the credit for this.

Prostitutes still skulked around back alleys in the Age of Reason, but now there were more of them than ever before. And some of them didn't even

---

* A sadist, by the way, is someone who gets a sexual thrill out of torturing people. But a Sadian (say the academics) is someone who spends his life studying de Sade.

** Including Byron in the Age of Reason may be a little unconventional because he is usually lumped in with the Romantics. But he was a product of the Enlightenment.

skulk. According to historian Vern Bullough, prostitution in England became "accepted as a fact of life, as something to be tolerated and accepted rather than abolished." In 1751 Parliament passed a law allowing for the prosecution of the owners of whorehouses, but few were ever charged. Bullough says the purpose of the law was to control prostitution, not to prohibit it.

I heard some people were upset when a couple of go-getters began publishing guides to the location of whorehouses, but the things sold like hotcakes.

In Paris, the Age of Reason could as well be known as the Age of the Bordello. If you didn't go to a bordello people thought there must be something wrong with you. Even priests regularly visited bordellos . . . enough priests to attract the interest of King Louis XV, who, for reasons known only to himself, kept track of the priests' comings and goings in a volume he called *Paris Nights,* though this was something of a misnomer. It's a well-established fact that people visited prostitutes during the day as well as the night.*[7]

---

* They still do, as anybody who's strolled down the Rue St. Denis in Paris can attest.

SEX: III

If people have a lot of woolly ideas about sex, it may be due in many cases not so much to misinformation as to a lack of information.

Take Greek literature. Everybody knows the Greeks wrote great literature. What people don't know is that the Greeks loved to write about sex. Aristophanes' plays, for example, are replete with sexual themes. But I bet your teachers never mentioned this. Nor do the standard encyclopedias. Aristophanes' play *Lysistrata,* for instance, is about women who refuse to have sex with their husbands until the men give up war. But as Harvey Einbinder pointed out in an exposé, the old *Encyclopaedia Britannica* merely noted that the play is about war. In the review of *Ecclesiazusae,* the *Britannica* neglected to say that the utopian society organized by women was organized in such a way that old women were given frequent opportunities to fornicate with young men. If a young man proposed to a young woman, for example, he first had to agree to have sex with an older woman.[8]

Diogenes is remembered as the Greek philosopher who lived in a tub and went around with a lantern looking for truth. Nobody ever mentions that he wrote plays advocating incest and cannibalism.*[9]

Or consider masturbation. What's usually left unsaid about masturbation is that the concern with it is relatively recent. Lawrence Stone reports that masturbation seems not to have aroused much attention in Europe until about the eighteenth century. And not until the nineteenth century did clergymen generally begin warning it could make you go blind or crazy. To be sure, clergymen had long denounced the practice. But given all the sins they had to worry about, this one didn't seem to rank too high. Even Calvinists, says Stone, "displayed only mild anxiety about the matter."

Another neglected area is the sex scandal. If too much is made of sex scandals in the present, too little is made of those in the past. Everybody reads Charles Dickens in school, for instance, but how many know he starred in an English sex scandal that nearly destroyed his career? What happened was this. Dickens fell in love with a young lady half his age named Ellen Ternan, set her up in a house next to his, then wrote the newspapers a letter denying he was having an affair with her, which was interesting, as nobody outside his little circle had ever suspected he was. Why'd he ever write the letter? To this day no one's ever figured that out.

A lot of celebrated Victorians had affairs, though little is heard about them. William Makepeace Thack-

---

* Diogenes, incidentally, didn't live in a tub. He lived in a huge clay pot turned on its side.

ery, behind his ill wife's back, had an affair with the wife of one of his friends. Lord Palmerston, at age eighty, fathered an illegitimate child.* Charles Parnell, the Irish leader, committed adultery with one Kitty O'Shea, leading to his downfall. Lloyd George, on a regular basis, had sex with his housemaids.[10]

And if not nearly enough is said about the sex scandals of the Victorians, clearly not enough is said about those of England's kings and queens. It's usually thought that in the old days at least they actually worked for a living. They did, too, but they also spent a lot of time in bed, though it wasn't always their own, and sometimes they preferred the bushes. The royal sex scandals of today have a great tradition behind them.

Burke's Peerage, as everyone knows, keeps track of Britain's royal lines. Less well known is that Burke's, as a kind of sideline, also apparently keeps a record of royal scandals. In 1991, during one of the royal controversies involving Prince Philip, the director of Burke's, Harold Brooks-Baker, disclosed in the *New York Times* the list of illegitimate children fathered by English monarchs over the last thousand years. Brooks-Baker reported that Henry I fathered 21 illegitimate children; Stephen, 3; Henry II, 2; Richard I, 1; John, 8; Edward I, 1; Edward II, 1; Edward III, 1; Edward IV, 2; Henry VIII, 1; Charles II, 14; James II, 6; George I, 4; George IV, 2; and William IV, 11.

---

* Disraeli, his political enemy, is said to have known about the birth but never made an issue of it. He feared if word got out that Palmerston—at age 80!—had had an affair, people would admire him more.

George II, of all England's fornicating monarchs, may deserve to be considered the most brazen. Every time he went out with another woman he told his wife about it. His wife's reaction? She made his first mistress one of her ladies-in-waiting. She shared his letters about his second mistress with Robert Walpole, George's prime minister.[11]

Four of England's kings (at least four) had homosexual lovers—William Rufus, Richard I (Richard Lion Heart), Edward II, and James I—but nobody mentions it. All you ever hear about Edward II, for instance, is that his wife ran away with another man and then came back with an army and overthrew him. The history books seldom say why she ran away.

Everybody knows of the scandals involving many of the French kings, but it's always made to seem like it was their own fault that they got into so much libidinous trouble. In fact it may have just been the way they were raised. Take Louis XIII. As a small child they let courtiers kiss his penis. And when he became curious about the female body, he was "allowed to poke his little fist up the vaginas" of his ladies-in-waiting.

Yet another neglected subject is infanticide. It's thought that in the old days when women got pregnant with an unwanted baby they either had an abortion or kept it. But more often than you might think, they murdered their babies. The practice ceased only in the eighteenth century. One of the reasons, says David Brion Davis, for the population explosion in western Europe in the eighteenth century was the "massive decline in infanticide."

Greeks and Romans also approved of infanticide, both Pliny and Seneca defending the practice. Edward Gibbon considered it "the prevailing and stubborn vice of antiquity."

Jews and Christians denounced infanticide and Constantine outlawed it (in A.D. 318). But infanticide continued on a huge scale in the Middle Ages. Only now people didn't go around killing their babies with their own hands. They let them die through exposure to the elements. It was considered unacceptable to bludgeon an infant to death; that was infanticide. But if you left a baby out in the open exposed to the elements and it died, that was "exposure," and hardly anybody complained. William Lecky, in his *History of European Morals,* wrote in 1869 that exposure "was practiced on a gigantic scale with absolute impunity, noticed by writers with most frigid indifference and, at least in the case of destitute parents, considered a very venial offence."

Since infanticide is a terrible crime, the English were especially careful about how they defined it. As late as the nineteenth century they were still grappling with the subject. What they finally decided was that it was infanticide if you killed the baby once it was completely outside the mother's body and breathing on its own. But if you bashed in the baby's head or slit its throat while part of it was still inside the mother, that was legal. In the language of the law, "it must be proved that the entire body of the child has actually been born into the world in a living state" for a crime to have been committed. The law remained on the books until its repeal in 1929.

In the eighteenth and nineteenth centuries, as people became much more civilized, they changed the method of getting rid of unwanted babies. Now they abandoned them on church doorsteps.

One of the big problems the churches faced was that they often didn't know when a baby had been left on their doorsteps. It sounds ridiculous, but a lot of babies died because mothers fearful of getting caught didn't dare ring the church doorbell when leaving a baby. Napoleon, though nobody remembers this, came up with the solution. He ordered hospitals to be equipped with a turntable, so mothers could leave the baby on the outside, ring the bell, and escape without detection. It was one of Napoleon's most successful reforms, only it worked too well. It became so easy to abandon a baby that soon it seemed almost every mother did. In the 1830s in France 32,000 babies a year were being abandoned.

Napoleon wasn't really to blame, though. Everywhere in Europe people were abandoning babies. In Spain in the 1830s 15,000 a year were being abandoned; in Italy, 33,000.[12]

About the history of homosexuality there are a number of myths. Having heard a lot about homosexuality in ancient Greece, for instance, the modern reader might be left with the impression that it was only the ancient Greeks who regarded homosexuality as socially acceptable. Actually, it was condoned or tolerated by ancient Celts, Germans, and Persians as well. But in all cases homosexuality was largely confined to the military elites. Its decline in Greece began around the fifth century B.C., but not because moralists decided it was bad. It declined because it was associated with the aristocracy and by the fifth century (the Age of Democracy), the aristocracy was on the run.

Another myth is that homosexuals have only recently begun coming out of the closet. Actually, though more gays are out of the closet today than ever before, large numbers of people began to identify themselves as gay hundreds of years ago, starting in the seventeenth century. It was then that the gay subculture first appeared, notably in London, possibly as a consequence of the birth of the modern British navy, which led to the city being crowded with sailors.

Englishmen who didn't want to go out in London's gay clubs could, by the eighteenth century, relax in similar clubs on the continent. Historians have identified so-called homoerotic clubs in France, Holland, Germany, and Italy. Gay Englishmen apparently felt there was less risk attending a gay club on the continent than one at home. One of the reasons the grand tour of Europe was considered so grand was because Englishmen found they could satisfy their sexual desires in Europe more easily than in England. This was true for gays as well as straights.

To be sure, homosexuality is more openly practiced today than it has been since ancient Greco-Roman times. But some of Europe's most important historical figures were openly gay. England's Richard Lion Heart and Edward II, by all accounts, were openly gay (at least in their court circles). So was France's Philip I. Philip even appointed his lover as a bishop. (Pope Urban II is said to have been aware of their relationship.)

Homosexuality has been regarded with suspicion in the West at least since the birth of Christ, but it hasn't always been regarded as dangerous. In a pioneering study, Yale University professor John Boswell found that in the first five hundred years of Christianity homosexuality was widely tolerated. Not until the sixth century A.D. did the Roman Empire flatly outlaw homosexual behavior, "even though Christianity had been the state religion for more than two centuries." In the early Middle Ages, says Professor Boswell, homosexuality was considered less offensive than adultery. Pope Saint Gregory III in the eighth century, for instance, punished priests more severely for going hunting than for engaging in homosexual acts. (Penance for a homosexual act lasted one year; for hunting, three.) The tradition of open and vehement hostility to homosexuality only began in the twelfth century. It was then that "homosexual behavior appears to have changed, in the eyes of the public, from the personal preference of a prosperous minority, satirized and celebrated in popular verse, to a dangerous, antisocial, and severely sinful aberration."

The Old Testament, incidentally, doesn't say homosexuality is evil. In Leviticus, the only place where homosexual acts are specifically referred to, the Bible says they are an "abomination." But abominations weren't evil. An abomination, says John Boswell, was simply "something which is ritually unclean for Jews, like eating pork or enjoying intercourse during menstruation."[13]

Interestingly, persecution of the Jews began at the same time as persecution of gays. This is a pattern that seemed to be followed closely through history. The Nazis, for instance, went after gays as well as Jews, killing more than 220,000.

It's sometimes said that there were a lot of homosexuals among the top Nazis, but there seems to have been just one: Ernst Röhm, the SA chief of staff. And Hitler got rid of him in 1934, on the Night of the Long Knives. Röhm seems to have lasted as long as he did only because he was useful in organizing the Brownshirt rabble into a strong fighting force.[14]

PART 5

GOD SAVE THE KING!

TRADITION AND ALL THAT

A DYSFUNCTIONAL FAMILY

RICHARD LION HEART

HENRY V

RICHARD III

GEORGE III

VICTORIA

EDWARD VIII

Crazy George III

TRADITION AND ALL THAT

The British do go on rather at length about tradition.

Take the monarchy. It is indeed as old as people say it is, but the pageantry associated with it isn't. Two of Britain's own historians* have proved that much of the pageantry the British find so dear was invented in the last hundred years or so. Consider Queen Victoria. At her coronation they didn't sing the national anthem, "God Save the Queen." The novelty shops didn't put ashtrays on sale with the queen's picture on them. And the clergymen presiding at the ceremony didn't wear fancy purple copes or colored stoles as they do nowadays.** To be sure, she got to ride around in a nice horse-drawn carriage, but observers commented snidely that hers wasn't terribly impressive. (Everybody liked the French ambassador's coach better; not until Edward VII did the monarchy finally obtain a decent gilded coach.)

---

* Eric Hobsbawm and David Cannadine.

** In the ecclesiastical world, a cope is a floor-length gown and a stole is a stole.

Victoria, I hasten to add, wasn't too interested in pageantry anyway. At her Golden Jubilee she refused to wear her crown and wouldn't wear her royal robes either. For forty years or so she even passed up the opportunity to open Parliament.

The British do pageantry so well now, everybody thinks they've been doing it the same way for a thousand years. But the British are just good at pretending they have.

Remember when Charles wed Lady Di? Everyone watching, I'm sure, thought what they were seeing—the royal phaeton majestically winding its way past well-behaved London crowds, the ceremonious parade of dignitaries, the clergymen dressed in splendid copes—was one of those classic British proceedings, rich in royal tradition. But it wasn't. They didn't start marrying princes in public until 1923.

Today the British do tradition so well hardly anyone can tell the difference between an ancient tradition and a new one. It's gotten so they even fool themselves.

But the truth is, except for a brief period in the fifteenth and sixteenth centuries, the British usually didn't do royal pageantry well. The problem was supposed to be in their genes or something. As Lord Robert Cecil explained in 1860: The aptitude for pageantry "is generally confined to the people of a southern climate and of non-Teutonic parentage." Times change. Today, people think pageantry *is* in British genes.

Whatever the cause of their problem with pageantry, they had a problem with it. The fact is they didn't even used to be too thrilled to throw a funeral. Until the nineteenth century, for instance, you could be a dowager queen and die and nobody seemed to care. They buried you in private. And it wasn't until the twentieth century that they put the body of those given public funerals on display at Westminster Abbey. It was first done in 1910 at the funeral of Edward VII.

And when they did start throwing funerals something always went wrong. In 1817 at the funeral of Princess Charlotte the undertakers got drunk. At the funeral of George IV his successor, William IV, "talked constantly and walked out early." When William's own funeral was held, the mourners "loitered, laughed, gossiped and sniggered within sight of the coffin."

I wish I could report that they knew how to do royal coronations better than funerals, but they didn't.

George III's coronation, for example, was a fiasco. First, it started late. Then, they discovered that they had forgotten the chairs for the king and queen to sit on. And then they realized they'd forgotten the sword of state. There was also a problem with a horse. It seems there was this horse that had been trained to walk backwards. The idea was that after the horse had been shown to the king it would gracefully back away so the king wouldn't have to stare at its ass. But, of course, the horse got a little mixed up and began walking backwards the moment it entered the hall and it kept walking backwards until it finally reached the king's table, ass first.

At George IV's coronation they still hadn't gotten things right. The king's costume was ridiculed for making him look too large ("indeed he was more like an elephant than a man"). His wife Caroline was blocked from entering the Abbey (at George's request). And professional boxers were employed to stop fights from breaking out among the guests.*

William IV, George's successor, didn't even want a coronation. Of course, they made him have one anyway. Of course, it was a disaster. This, however, was nobody's fault but William's. He conducted the coronation on the cheap and it showed. Afterward, it became known as the "Half-Crownation." Even Queen Victoria's coronation went off badly. The Archbishop of Canterbury "put the ring on a finger which was too big for it." The presiding clergyman lost his place during the service. The choir sang badly. And the trainbearers talked when there was supposed to be silence.

Why they got into the business of manufacturing traditions is easy to answer. They saw a need and filled it. Millions of people around the world find it reassuring to know that there are still a few human beings who get the opportunity every now and then to put on colorful stage costumes and ride around in horse-drawn carriages.

The idea of putting the royal family on parade came just in the nick of time. At the end of the nineteenth century it was beginning to seem that the monarchy didn't really have a purpose. Now everyone could see that it truly did.[1]

---

* George and Caroline had always hated each other. Theirs had been an arranged marriage. Of Caroline, George once said: "I had rather see toads and vipers crawling over my victuals than sit at the same table as her." She died within weeks of his coronation.

A DYSFUNCTIONAL FAMILY

That the royal family is regarded today as newly dysfunctional is unfair. English royal families have a history of dysfunctionalism. All of the Georges hated their fathers. Fifteen of the kings fathered children out of wedlock. Brother fought brother (Henry I v. Robert). Son fought father (Richard Lion Heart v. Henry II). One time, a wife (Isabella) helped depose her husband (Edward II) so her lover (Roger de Mortimer) could take his place. And, of course, there's the case of Henry VIII. But too much fun is made of Henry. So he had a little trouble finding the right spouse, who doesn't?

Speaking of the monarchy, this is probably the place to point out it wasn't too popular much of the time. For most of the nineteenth century, for example, English royalty "almost without exception"* was "viewed with indifference or hostility." George III wasn't too popular, of course, but you

* I'm quoting David Cannadine, the British historian referred to earlier.

probably knew that.* But neither was his son, George IV. George IV had the problem common to a lot of unpopular people—nobody liked him. People disliked him so much they didn't even pretend on his death that they liked him. When he died the *Times* of London editorialized: "There never was an individual less regretted by his fellow creatures than this deceased king." Victoria was more popular. But even she faced sustained criticism. So much criticism, she was afraid to appear in public at her Golden Jubilee. Some people even dared question what she did with her money.

In the nineteenth century, the people the British admired the most were military leaders. Nelson and Wellington, for example, were both more popular than the royals and were given better funerals. Most revealing of all, perhaps, their faces appeared on more plates and ashtrays and knickknacks than even the king's.[2]

---

* Actually, he wasn't as bad as they say. See p. 100.

RICHARD LION HEART

Richard Lion Heart was a big robust man who went around killing people. The English loved him.

He had a good father (Henry II) and a bad mother (Eleanor of Aquitaine). Naturally, he took after his mother, who was notorious for making war on her own family. Like the time she called out the army on her husband.*

Richard is remembered as the swashbuckling hero who volunteered for the Crusades and made England proud. Of course, he never did get back Jerusalem, but what did that matter?

---

* Why Henry married her is clear enough. She was the richest woman in the world, owner of Aquitaine, a huge region in southwest France. By marrying Eleanor, Henry (who already owned Normandy and Anjou) came into the possession of more of France than the king of France, Louis VII. This circumstance led to a great deal of ill will between the two monarchs and ended in war. Eleanor, incidentally, had been married to both. She married Henry after Louis dropped her.

Some have said he didn't behave in a Christian way over in the Holy Lands. The critics point to the time he slaughtered three thousand Moslem captives, including their wives and children. But the English didn't seem to mind.*

A little later, in hopes of creating a joint Christian-Moslem kingdom, he tried marrying his sister off to Saladin's brother. (Saladin was the leader of the Moslems.) But Saladin's brother, for some reason, didn't seem too interested. Neither was Richard's sister.

To his people back home, of course, Richard remained a great hero. Even if, in his absence, crime was up, the realm was nearly broke, and the nobility was out of control.

Eventually, of course, the time to go home finally arrived, as Richard had done about as much good as any man could over in the Holy Lands. Unfortunately, though he was one of the most ingenious strategists of all time, on his way home he was captured by Leopold of Austria and imprisoned for a year. His people had to bail him out.

Be that as it may, the English loved him and welcomed him home.

Richard wasn't cut out to govern, though, so he went back to fighting as soon as he could. In 1199, however, during an attack on a castle where

---

* So he "quarreled now and then," as Lady Callcott wrote in her children's history of England. Deep down he was "really good-natured." (*Little Arthur's History of England* [1835; rpt. 1981], chapter 20.)

some secret treasure was supposedly hidden, he died. An arrow got him.

Just why the English loved Richard I've never figured out. He spent most of his time abroad. He never learned English. And he "bled the land white to finance the Crusades."[3]

HENRY V

King Henry V is well known as the legendary hero who defeated the French at Agincourt, in the "greatest battle in English history."

Not too many people know where Agincourt is, but that hardly seems to matter.*

Shakespeare is responsible for the belief that Henry ran around with robbers and highwaymen as a youth, but Shakespeare seems to have gotten this part wrong. As far as anyone can tell, young Henry was a model citizen.

The story told about Henry and the justice—that Chief Justice William Gascoigne put Henry in jail when Henry struck him on the head—is also without foundation. It was supposed to have happened when Henry was the Prince of Wales and is told for the ending. After Henry was in jail a few days he is said to have calmed down and seen the wisdom in

---

* *Webster's Geographical Dictionary* tells us it's "33 m. WNW of Arras." That's helpful, isn't it? So where's Arras? It's "25 m. SSW of Lille." Where's Lille? It's "130 m. NNE of Paris."

the justice's action, proving that Henry was fair and honest and all that. Henry is even said to have promised the justice that he would never "behave so ill again." But the whole thing is apocryphal.

Upon his death the crown went to his son, Henry VI. Before long Henry VI lost all the territory in France that his father had gained. Which is why in France they speak French instead of English.[4]

RICHARD III

King Richard III's reputation as the meanest, vilest, ugliest monarch in English history is so well established that not even Hollywood stars Helen Hayes and Tallulah Bankhead—founders of the "Friends of Richard III" society—were able to change many minds about him. But *was* he mean, vile, and ugly?

Mean, he was. I think we can all agree on that, even allowing as how standards of meanness have changed over the years. I believe even in the fifteenth century they considered it mean to cut off a person's head.

Vile? Here we come to a more complicated question. We have all heard that he killed his two nephews, "the sweet little princes," which would seem to be a pretty clear example of vileness. But no one knows if he did it or he didn't. Shakespeare says he did, but Shakespeare just wanted to make Richard's successor, Henry Tudor, look good by comparison.

A lowly no-goodnik named James Tyrell is said to have confessed twenty years later that he'd committed the crime at Richard's behest. But I think it's a little unusual that nobody happened to mention the confession until after Tyrell had died.

Tyrell, to be sure, was just the kind of man who'd figure in the murder of a couple of innocent kids, seeing as how he was guilty of at least one other murder that we know of. But nothing links him to the deaths of the little princes except his so-called confession.

Ugly? Shakespeare says Richard was so ugly dogs barked at him when he passed by. But his portraits show he was almost handsome. That he was a hunchback with a withered arm is poppycock.

Shakespeare, I know, tells a different story. But Shakespeare's history was slanted.[5]

GEORGE III

In British school books George III is usually portrayed as an intelligent and capable monarch and a good family man. Naturally, we Americans feel a little differently about him.

But was he the great devil-monarch we've made him out to be? Apparently not. *The Founding Fathers actually misled us.* George was not the only one hostile to Americans. The British people were against us, too. Studies show that the House of Commons was a hotbed of anti-Americanism.

That he is regarded as a tyrant is now believed inaccurate. George III was a strict constitutionalist. In his whole long reign (sixty years long), he is said to have carefully respected the prerogatives of Parliament. The belief that he packed the legislature with favorites and used bribes to get his way is without foundation. His enemies said he did but now we know he didn't.

Question: Who said, "The pride, the glory of Britain and the direct end of its constitution is political liberty"? John Locke? No. It was that well-known tyrant, George III.

Maybe most remarkable of all was that he was a devoted father and a faithful husband. His whole life he never strayed once, we are told. This must be almost some kind of record for a British monarch. And to think that he remained true to his wife at a time when everybody else seemed to have been uncontrollably sexual is, well, as I said, remarkable.*

To round out the portrait, maybe it's worth mentioning his many interests. He loved music. He loved books. He loved astronomy. And he loved clocks. But it may be he loved them all a little too much. He didn't just enjoy listening to Handel every now and then, he had to listen to him, day in, day out, virtually every day of his reign. It has seemed to some a bit obsessive. And he didn't just like to buy a book now and then, he had to buy them by the tens of thousands, until he had collected more books than any other monarch in English history. And he didn't just enjoy gazing at the stars, he had to build the largest observatory in the world so he could see them more clearly than anybody else ever had. And he couldn't just make do with one or two clocks, he had to have dozens of them, a clock for each and every room in his castles, and each and every one of them always ACCURATE! (He hired an expert clockmaker to keep all his clocks exactly on time. Under George, it was a full-time job.) And, of course, since he

* This was the Age of Sex, after all, remember? (If this doesn't sound familiar it means you skipped, for some insane reason, the section on "Sex." Before proceeding further see pp. 69–84 immediately!)

liked clocks, he had to have a wristwatch, and it had to be the most perfect wristwatch ever produced, so accurate it automatically compensated for changes in temperature; "hot or cold, the King had the right time."

Some think his obsessiveness raises questions about his sanity, which has always seemed somewhat in question anyhow. After all, there aren't many monarchs who got so out of control they had to be tied up in a straitjacket as George was.

But he probably wasn't crazy. Medical experts now believe he was suffering from a hereditary illness known as porphyria, whose symptoms are inordinate restlessness, delirium, and rashes, symptoms George displayed whenever he seemed to go insane. It cannot be proven that he had porphyria. But it showed up in several generations of his descendants.

I'm sure the experts are right. But there's still something that bothers me. He seems to have carried his hatred against his father further than a sane person would. His whole life George III refused to sleep in any castle his father had slept in. He had any number of perfectly good castles, but if his father had lived in them, he didn't want to have anything to do with them. Take Hampton Court. It's one of the prettiest castles the English ever built, full of fine old furniture and fancy drapes. But because it had been used by his father George hated it. He hated it so much that when the castle caught on fire, he hoped it would burn down and was annoyed when it didn't.[6]

VICTORIA

Queen Victoria knew students would have a hard time remembering all of the names of the English monarchs, so she conveniently named the era in which she lived after herself to make things easier. But she wasn't a typical Victorian.

She wasn't stuffy and she wasn't reserved. She liked to drink and she encouraged others to. And she tolerated open drunkenness in her court. Her favorite drink: claret diluted with whiskey.

When her husband Albert died she mourned in the conventional manner, dressing in black and refusing to be seen in public. But she remained in mourning a little longer than was customary even for a Victorian: some twenty years.

She did not remain single, however. A few years after Albert died she got a companion, John Brown, an erstwhile Scottish servant who lived in the palace. What they had in common is unclear. He grew up poor and spoke bad English, but he was well-built and handsome; and like Victoria, he enjoyed drinking. After a while gossips began to refer to the queen as Mrs. John Brown.

Whether they ever had sex is unknown, but after her death the royal family secretly purchased 300 letters Victoria wrote to her physician. The letters were said to include compromising references to Mr. Brown.[7]

EDWARD VIII

Edward VIII is the king who gave up his crown for the love of a woman. Unfortunately, as he wasn't the hero type and she wasn't either, it's not as good a story as it should be.

As a young man, he had been different from most princes in that it appeared he had a social conscience. During World War I he visited the wounded in the infirmaries and even spent a night or two out in the trenches. But as this kind of thing wasn't very exciting, he gradually gave it up and devoted himself to dandyism and girls, for which he was much better suited by nature. His years as a randy fop, unfortunately, lasted a little longer than they probably should have.

His private secretary, Tommy Lascelles, later said some nasty things about him, but it may be that Lascelles was prejudiced. Edward had never heard of *Jane Eyre* until Lascelles told him about the book. After that Lascelles never gave Edward a chance.

Most people who knew Edward weren't too impressed with him. Prime Minister Stanley Baldwin hoped Edward would break his neck the next time he went racing.

Edward's father, King George V, predicted that Edward wouldn't last. "After I am dead," said the king, "the boy will ruin himself in twelve months." He was wrong. It took only ten months.

Eventually, Edward settled on one woman, Wallis Simpson. Unfortunately, she probably wasn't the best choice, as she had a habit of marrying men and then divorcing them. Also, she was still married to her second husband when she started going out with Edward.

Whether she was a social-climbing parvenu, I don't know. But she named her childhood dolls after Mrs. Astor and Mrs. Vanderbilt.

To his credit, Edward did everything he could to prevent a scandal, especially after he became king. When she and he registered together at the same hotel in Vienna, for example, he discreetly signed himself in as the Duke of Lancaster.

The gossip is that they slept together before they got married, but when a writer claimed in a book that they had, Edward sued him and won. I might add that the judge was a personal friend of Edward's. But I'm sure that had nothing to do with the outcome of the case.

The English people, interestingly, were the last in the whole world to know about the scandal, as news of Mrs. Simpson was banned from the local papers and literally cut out of the foreign ones sent into the country. A member of Parliament did ask one day about the holes in the papers, but nobody else seemed to wonder about them.

When the scandal finally did break in the English

press it was Mrs. Simpson's past divorce that seemed to bother people the most. But I think the British took a rather extreme view of divorce. Lord Halifax, a High Churchman, even once debated whether it was proper to be seated next to a divorced woman at dinner. It was all right to sit next to someone who'd committed adultery, people were sure. But to sit next to someone who'd committed divorce, well, that was pushing the boundaries, wasn't it?

In the end, of course, Edward was forced to abdicate. But this wasn't the tragedy it's made out to be. For one thing, Edward hadn't enjoyed being king anyway. He didn't like reading cabinet papers and he didn't like the ribbon-cutting stuff.*

For another, he wasn't exactly the right king for the times. He liked Germany under Hitler a bit more than was desirable in an English king. Of course, England wasn't yet at war with Germany, so his behavior wasn't treasonous. But when he publicly approved of Hitler's decision to remilitarize the Rhineland he got a few good hard looks.

It's also reported that on one occasion he held up the reception line at a gathering of diplomats so he could talk with the German foreign minister. This made all the other diplomats jealous and caused a big row.

He also expressed a bit more admiration for Mussolini than was wise for a man in his position.

Apparently, Edward never once thought he'd done anything wrong, though he must have wondered at times why Foreign Secretary Anthony

---

* He did like the traveling and the castles, though.

Eden was turning gray so fast. But Edward wasn't one to worry about other people's problems.

Edward was very old-fashioned in a way. He thought it was his job to run English foreign policy. Apparently in school he'd skipped over the lesson where the king gave up this duty.

Between fights with Anthony Eden, Edward would go on little trips. But even these would sometimes land him in the soup. Like the time he planned to sail his royal yacht into Venice. Edward saw it as a nice way to spend a few carefree days under the Italian sun. But as Mussolini had just invaded Abyssinia, and as England had condemned the invasion, and as the king was supposed to set a good example, Eden got a little upset.*

Edward professed not to know what on earth was bothering Eden.

At one point Eden confessed to friends that Edward would have to abdicate if he didn't quit interfering in foreign policy. Things had gotten that bad! And then, along came the Simpson crisis.

Some have wondered why the cabinet did not spring to Edward's defense when the scandal broke. Me, I don't wonder.

Edward was succeeded by his brother Albert, who went by the name George VI. Don't ask why Albert didn't become King Albert. Even the English aren't too sure. He just decided he'd rather be

---

* You remember the Abyssinia crisis. Mussolini invaded Abyssinia (Ethiopia). Haile Selassie gave a wonderful speech condemning the invasion, which moved the entire world to tears. Then people forgot all about it.

called King George. Now, there had already been five kings named George. But Albert thought there was always room on the royal genealogy charts for one more.

Even the royal family must have had trouble sometimes remembering who was who. Consider, for a moment, just the sons of George V. There were the two Georges: the George who was really Albert and the George who was really George. Then, of course, there was Edward, who was known in the family as David. Finally, there were Henry and John. Happily, Henry went by Henry and John by John.

Why Edward was known in the family as David is simple. It was one of his seven Christian names and it was the one they called him from birth and liked best. His full name, in case you're interested, was: Edward Albert Christian George Andrew Patrick David.

What came of Edward? Edward remained true to himself: confused and slightly ridiculous.

Among his favorite activities was giving inane interviews. Worth mentioning is the one he gave to Fulton Oursler in 1940, at the height of the Battle of Britain. In the course of the interview the ex-king of England proclaimed that Hitler was a great man, that Germans should be delighted to have such a leader, and that Britain would be dumb to fight Hitler as Hitler was unbeatable.

He then went on to explain his idea of how to settle the conflict. "It sounds very silly to put it this way," he said, "but the time is coming when somebody has got to say, you two boys have fought long enough and now you have to kiss and make up."

What he really had in mind, no one knows. I doubt even Edward himself was all too sure. Edward was never too clear about anything. But it apparently involved him teaming up with Franklin Roosevelt to make an end-run around the current British government. As Edward explained it: (1) Roosevelt would call for peace. (2) Edward would promptly issue an anouncement saying he, too, favored peace. And (3) there would be a revolution in England and peace would break out.

Left unsaid was whether Edward expected to be reinstalled as king. I think, though, that was kind of the general idea.

As a husband, Edward proved better than anyone could expect. He and Mrs. Simpson remained married to the end.

He didn't do so well with his relatives, though. His story was that they treated him rotten after he abdicated. Their story is that he deserved to be treated rotten.

With relatives, of course, there are always bound to be problems. But I think it was a mistake for Edward to lie about his finances during the abdication crisis so he'd get more money.[8]

PART 6

"THIS SCEPTER'D ISLE"

MAGNA CARTA

STAR CHAMBER

DEFEAT OF THE SPANISH ARMADA

CAPTAIN KIDD

BLACK HOLE OF CALCUTTA

WILLIAM BLIGH

HORATIO NELSON

LAWRENCE OF ARABIA

OF THINGS OLD

OF KILTS AND BAGPIPES

Magna Carta

MAGNA CARTA

Magna Carta is truly a remarkable document. I read that at the 1939 New York World's Fair it drew a crowd of ten million over just six months. Nine out of ten Americans believe it's a mortal sin to wait five minutes to see a doctor. And here were TEN MILLION willing to wait hours on line for the opportunity to spend one nanosecond in front of a torn and soiled parchment they couldn't even read.

As with most famous documents, though, there is a little disagreement about it. Some people say it's the "fountain of our liberty." Others say it isn't.

After seven hundred and fifty years or so you'd think maybe the experts could have decided by now what to make of the document, but they haven't.

All I know is, if it's the "fountain of our liberty" we're in trouble.

Remember the heinous medieval practice of trial by combat? Under Magna Carta it was legal. Trial by ordeal? It was also legal. (In a trial by ordeal the accused was allowed to prove his innocence by surviving a dunk in a vat of boiling tar.)

Trial by jury? People say Magna Carta provided for trial by jury, but it didn't. In 1215 in England they didn't have jury trials. Suspects didn't have the right to cross-examine witnesses, exclude hearsay evidence, produce a defense, or even wear a sack over their heads on their way to and from court.

How about the right to be tried by a jury of one's own peers? This indeed is a right everybody ought to have and you can find it in Magna Carta just as everybody thinks. It is one of several important rights to be found in the document. The catch is, only free people were allowed to exercise the new rights listed in Magna Carta, and in 1215 only a small number of Englishmen were free. Five-sixths were serfs.

So who really benefited from Magna Carta? England's barons. All the fuss about Magna Carta is about the new rights they won from the king for their own protection. But Magna Carta didn't give the average Englishman one more right than he'd had before.

Now, just for the sake of argument, let's say it could be proved that Magna Carta curtailed the power of the English monarchy, wouldn't that have been worth something to the average Englishman? The disappointing answer is that it wouldn't have. For the average Englishman in 1215 wasn't oppressed by the monarchy. He was oppressed by his baronial lord.

Magna Carta, anyway, didn't prove much of a check on the monarchy. It was *after* Magna Carta that England got its first true tyrant kings.

So why, if all this is true, do we celebrate Magna Carta today? It is because several hundred years ago a bright Englishman by the name of Sir Edward Coke decided to put one over on us all. He announced one day that something called Magna Carta, which he said he'd found sitting on an old dusty library shelf, gave Englishmen rights the monarch couldn't take away. And that was that. From that day forward the English felt all their rights and liberties could be traced to that one document, a document nobody had ever even heard of before.*

After the discovery of Magna Carta no English king could spit on the sidewalk without somebody jumping up and down and screaming, "Magna Carta. Magna Carta. Watch it, fella." It took all the fun out of being king.

King John, by the way, didn't really sign the Magna Carta. He had his royal seal pressed in wax on the document. All the Hollywood movies which show him signing it are wrong. A king did not deign to sign anything. (Many didn't know how.)

That he deserves to be remembered as a bad king is not in dispute. At issue is whether he was simply a bad king or a really bad king. He lost Normandy in a failed war in which he left his soldiers on the field while he fled for safety. He apparently murdered his cousin Arthur. He kept Arthur's sister locked up in a prison cell for forty years. He stole another man's bride for himself. He took hostages from his barons to guarantee their fidelity and then either ransomed or killed the hostages. And just for the fun of it, he locked up the wife and

---

* Shakespeare, in his play about King John, doesn't even mention it.

child of an erstwhile friend and starved them to death.

On the other hand, he was said to be a very efficient administrator, one of the best England's ever had.

What bothered the barons most about him was his position on rents. It was your classic landlord/tenant dispute. He wanted to raise their rents. They didn't want him to.

Incidentally, the barons, when they organized against John, called themselves the "Army of God." John's army was called John's army. It was an unequal match from the start.[1]

STAR CHAMBER

Why was it called the Star Chamber? No, it was not because "star" criminals were tried there. It was called the Star Chamber because gilded stars decorated the courtroom's ceiling. Incidentally, the name Star Chamber technically did not refer to the courtroom where the trials were held but to the building as a whole. It took the English a couple more hundred years to decide that a chamber was a room, not a building.

Whether a defendant was treated fairly or not by the court depended on who happened to be king at the time. Under good kings the court could be quite fair. Under bad kings, well, I'd rather not say, but a defendant could find himself in quite a pickle.

As courts went in those days, though, it was rather better than most, especially if you were poor. In the opinion of historians, in fact, it was the only court in England where the poor and the rich met on equal terms.

Better yet, it was one of the few courts in England where they didn't squeeze your head in a vise, tie iron wire round your private parts, stretch you on the rack, or make you watch repeat episodes of "The Six Million Dollar Man." They never sentenced anybody to death even. If you needed punishment they let you off with a fine or a jail sentence, though every now and then, just to show they were serious, they'd chop a person's ear off.

So why do people think it was some kind of Chamber of Horrors? The usual explanation is that under the bad Stuart kings it was used so often to enforce unpopular royal edicts that people plain just got crazy about it. And I'm sure there's something to this. But I think another reason is that the rich folks weren't happy with a court where they were on equal terms with the poor. And as they wrote the first histories of the institution, their prejudice got some publicity.

You might think historians would have cleared this one up long ago, but medieval records are in such bad shape they couldn't make out who did what when until just recently. As an example of the confusion, the records do not clearly distinguish the proceedings of the Star Chamber—where torture was not allowed—from those of the king's privy council—where torture was allowed.[2]

DEFEAT OF THE SPANISH ARMADA

Against all odds, in 1588 the English defeated the greatest navy in the world, the Spanish Armada.

How did they achieve this great victory? (Pick one.)

The Spanish were boobs.

This helped. Try as they might, the Spanish couldn't seem to do anything right. As an example, when they outfitted the ships with gun carriages, they put in the kind that cannot easily be pulled back from the gun port, making it impossible to reload the guns during battle. This was a mistake.

It was also a mistake, I think, for the Spanish to give the English advance warning of the attack, seeing as how it was supposed to be a secret and all. Of course, the Spanish didn't mean to. They thought they could trust the College of Cardinals with the information. How could they know there'd be a snitch among them?

The Spanish had a stupid plan.

This also helped. The main problem with the plan was it couldn't possibly work. What the Spanish counted on was that their two fleets, one from the south, the other from the east, would arrive at the agreed-upon point of rendezvous at precisely the same moment. But, of course, as the telephone hadn't yet been invented, there was no way in hell this was going to happen. By the time the fleet from the east arrived—two days late—the game was up.

A surprise storm wiped out a third of the Spanish force.

The English insist the storm arrived *after* they had already defeated the Armada, but this is in dispute.

The English had God on their side.

Perhaps. But it was the *Spanish* who were outnumbered. The English had 197 ships; the Spanish 130. Spain's ships were bigger, but in the narrow English Channel big ships were difficult to maneuver.

The English navy was well prepared.

Actually, no, not really. The defense of the country was left in the hands of a malnourished, ill-trained, and badly paid force of sailors, many of whom were so neglected that by the time the

Spanish arrived they weren't too interested in fighting.

Queen Elizabeth offered sterling leadership.

To be sure, she was brave and true and all that, but it was mainly because of her that the navy was run on the cheap.

William Burghley, her treasurer, deserves some of the blame, though, as he kept miscalculating the funds at her disposal, necessitating repeated cuts in the navy's budget. His problem, says historian Lawrence Stone, was that he only knew how to count in Roman numerals. This made for a lot of errors. Try adding XVIII plus CXI plus VIII yourself. Not easy, is it?

King Philip of Spain was a bad man.

Philip *was* peeved that Elizabeth had rejected his hand in marriage, but he didn't decide to fight her out of personal pique. She brought on the attack by her own foolishness. In the preceding three years she'd raided the Spanish Main, authorized Sir Francis Drake to take Spanish booty, and in 1585 promised to come to the defense of the Dutch provinces in rebellion against Spain.

A famous story about the Spanish Armada worth mentioning is that Sir Francis Drake played bowls on the lawn at Plymouth on the eve of the invasion. Though the Spanish fleet supposedly had already been sighted in the English Channel, Drake is said to have remarked, "Let us play out our match.

There will be plenty of time to win the game and beat the Spaniards, too."

Fascinating story. No evidence.

Finally, I suppose you heard that after the Armada was defeated Spain slipped into immediate decline. But this isn't exactly true. There is every reason to believe, as most historians do, that Spain was stronger after the defeat than before. Between 1588 and 1603, Spain recovered more treasure from the colonies than in any other prior fifteen-year period. And the navy, after some revamping, emerged leaner and more effective.[3]

CAPTAIN KIDD

Captain William Kidd was hanged on May 23, 1701. Then the rope broke, and they had to hang him again. His whole life was like that—one crazy thing after another.

Take this business about him being a pirate. It's topsy-turvy. There Kidd was, trying quietly to do his job as a *catcher* of pirates, when suddenly one day somebody accused him of *being* one.

Kidd wasn't even the pirate type. I don't know what the hell he looked like, but he was wealthy, had a nice little family, and went to church. Charles Laughton he wasn't. He didn't even growl.

Seeing as how we all think he was British through and through, I should mention that he lived in New York. The reason nobody knows this is that New York kindly let the British claim him after he got into trouble.

Kidd, if anything, was a little too good. I have in mind the time he went back to England to complain to the authorities about the way the British governor of New York had rigged an election. To me this sounds a bit much. I mean it's not like he could jump on the Concorde, do his business, and get back in time for lunch. He had to go by BOAT!

But the British were impressed, and soon after they made him an official pirate catcher. Unfortunately, as a pirate catcher, Kidd didn't do too well. In fact, in more than a year on the high seas, he managed to catch, to be precise, none.

This was something of a feat, as there were more pirates in those days than there were normal people.

After a while, unfortunately, tired of chasing pirates and not catching them, he himself started raiding ships and stealing their cargo. Which, on the face of it, looks pretty bad. But in the old days this was legal if you had a permit, and Kidd had one.

The trouble was he was only supposed to raid French ships and the two ships he raided were not French. Kidd's excuse for the raids was that the ships were traveling under French safe-conduct passes, which, if true, would have been somewhat exculpatory. But when he was asked at his trial to produce the passes, he couldn't.

Which brings me to: Kidd's trial.

As this was England—*home of MAGNA CARTA!*—Kidd's trial was, of course, fair. There were judges, lawyers, a jury and everything. The state even saw fit to give Kidd money so he could hire a couple of attorneys to help with his defense.

On the down side, he wasn't allowed to consult with his attorneys until the morning of the trial and on the second day they didn't even show up. Which was somewhat hurtful to his defense, as it was a two-day trial.

But what really damaged his cause, as I mentioned above, was his failure to produce the French passes. Which brings me to the moment you've all been waiting for, the moment it becomes clear an outrageous injustice has been done. It turns out the passes Kidd needed so badly were hidden in the prosecutor's drawer.

The prosecutor, of course, had denied knowing anything about the passes, but after the trial they mysteriously turned up. Today they can be seen on display in the London Public Records Office.

Kidd himself was not entirely blameless in the whole affair, of course, as nobody ever is. You may have heard, for instance, that he had hobnobbed with pirates on Madagascar, which is, I grant, interesting behavior for someone supposedly out catching pirates. But I suppose you and I would have done the same thing, too, if we'd been stuck on Madagascar, since the place was crawling with pirates, many of whom didn't think too highly of pirate catchers.

To give him his due, Kidd, upon sailing into port, actually tried to make a couple of arrests, but as his crew refused to follow his orders, and as they were in general mutiny, and as he was outnumbered, he decided not to do so.

You may have also heard that he'd struck one of his own sailors over the head with a bucket and killed him. This is also true. But as the sailor was refusing to follow orders, you can maybe understand why he did it. At worst, he was guilty of manslaughter, but he was charged with murder.

Why, if Kidd was not a pirate, has he been made out to be one? Why has he gone down in history as one of the baddest men who ever lived? It was because of politics. It had been the Whigs who'd sponsored his expedition, and when it didn't go too well, the Tories dragged him into court to make the Whigs look bad. Poor Kidd was just a scapegoat.[4]

BLACK HOLE OF CALCUTTA

This is, depending on whose account you read, either:

(a) a miserable little hole into which mean Hindus shoved 146 poor innocent English, Dutch, and Portuguese prisoners one hot summer night in 1756, a hole so cramped with people that by morning only 23 had survived; or

(b) a miserable hoax invented by mean British imperialists to justify a crackdown on poor, innocent Hindus then in rebellion against British/Moslem tyrants.

And which view is correct? I haven't any idea.[5]

WILLIAM BLIGH

William Bligh's crime was not that he was too harsh, but that he was too lenient. The movie got things backwards.

The reason his men mutinied on the high seas is because he'd let them roam free as birds on Tahiti for five months. After that they just couldn't seem to settle down anymore.

Admittedly, it was a stupid thing Bligh did. He should have known better than to let his men get a taste of the good life. It's just plain not good for people.

The story that he was the meanest, orneriest low-down sea dog in history was made up by the men who mutinied and got caught. It was given credence by the public, however, because at the men's trial the government didn't call Bligh in to contradict their accounts. And after the trial Fletcher Christian's family published a bogus transcript of the proceedings that led people to believe the claims against Bligh had been proven.

Christian, the dashing young leader of the rebels, also made a more attractive hero than Bligh, though Bligh was the genuine article. After the mutiny, Bligh and his small band of loyalists were put aboard a small boat and cast adrift thousands of miles from the nearest shore. But Bligh miraculously sailed the boat to safety.

What about Bligh's swearing? It's true. Bligh swore quite a bit. But historians say "his cursing was no worse than that of other commanders."

The complaint that he was a harsh disciplinarian is unfounded. He flogged only seven men the whole voyage, and they are said to have deserved it. One had tried to desert and three had mistreated native women. The others had just plain acted up.

Bligh, in fact, was a considerate commander. In the rain he'd give up his private cabin for sailors who were wet. And so the men wouldn't get bored he brought a blind fiddler on the trip.

What came of Christian? Some say he and his men ended up on Pitcairn Island. Others insist they were shipwrecked. Still others say Christian slipped back into England somehow and was protected by friends.

And Bligh? He was made an admiral. Then the authorities put him in charge of Australia. And, unbelievably, people mutinied against him yet again.[6]

HORATIO NELSON

Horatio Nelson, the man responsible for the victory of the English over the French in the Battle of Trafalgar, is remembered as the ultimate hero. He wasn't. Though he was brave and true and all the rest, he was a notorious adulterer. For years he carried on in public and private with Lady Hamilton, the wife of a minor British ambassador. They even had a child, Emma, whom Nelson admitted siring. Biographers in the Victorian era blamed Lady Hamilton for the liaison, claiming she'd wickedly seduced Nelson. But Nelson obviously should be held equally responsible.

Nelson was the naive type, though. Only someone naive could invite his mistress in to live with his wife and think things would work out well.

The phrase with which he is most associated, England Expects That Every Man Will Do His Duty, which he used at Trafalgar to galvanize his sailors, is thought to have captured the essence of English patriotism. It was, however, supposed to read *Nelson* Expects That Every Man Will Do His Duty. The change was made only because they didn't have a flag that spelled out the name Nelson.

It was at the Battle of Trafalgar that Nelson lost his life. But it wasn't because he was brave that he died. He insisted on parading up and down the deck of his ship dressed in full admiral's regalia, where he could easily be picked off by an enemy sharpshooter—and was.[7]

LAWRENCE OF ARABIA

Lawrence of Arabia was many things: A courageous soldier. A great writer. A handsome man. BUT HE WAS NOT THE BEST FRIEND THE ARABS EVER HAD.

I'm afraid we have all been misled. The truth is, Lawrence sold them out. In World War I he did not really champion the Arab Revolt, he only pretended to. From his letters we now know he never intended to help free the Arabs and he never tried to help them unite. As he told his superiors in London, he believed it was best if the Arabs remained "a tissue of small jealous principalities."

The movie, I know, indicates Lawrence *did* promote the Arab cause. Well, the movie is wrong.

Why did Lawrence pretend to support the Arab Revolt? It was because in World War I, as he put it, "Arab help was necessary to our cheap and speedy victory in the East." "I could see that if we won the war the promises made to the Arabs were dead paper," he wrote in a letter to a friend. "Had I been an honourable adviser I would have sent my men home, and not let them risk their lives for such stuff. Yet the Arab inspiration was our main tool in winning the Eastern war."

Now about the controversy involving his sexuality. Was he a homosexual or wasn't he?*

By his own admission, Lawrence never had sex with a woman and he repeatedly said he never wanted to. As he explained to a friend, intercourse is "dirty." Besides, "it's all over in ten minutes." Lawrence, again by his own admission, indicated he *had* engaged in sex with a man, but it was only once and it wasn't voluntary. It happened during the Arab Revolt when Lawrence was taken prisoner by the Governor of Deraa and tortured and buggered. As he confessed in a letter: "For fear of being hurt, or rather to earn five minutes respite from a pain which drove me mad, I gave away the only possession we are born with—bodily integrity."

There is no evidence Lawrence ever had intercourse with another person on any other occasion. But during the last ten years of his life Lawrence repeatedly gave in to a masochistic desire to be whipped.

Why he liked to be whipped is one of those things nobody will ever know for sure. Probably Lawrence himself didn't know. It may have been for the thrill. Or it may have been because his mother had whipped him as a boy. Or it may have been punishment for giving in to buggery in Deraa. But it's worth pointing out, he always insisted on being whipped by a man.[8]

---

* Were you wondering if I was going to get to this?

OF THINGS OLD

I come now to the widely held belief that all British buildings are old. This is simply not true. Many British buildings just look old. The Houses of Parliament? They were built only in 1860. The Gothic support arches are fake. The building was constructed the modern way, with a steel frame. Stephen's Chapel, where Parliament used to meet, that was old. But Stephen's Chapel burned down in a terrible fire in 1823.

Big Ben? Not old either. It was built when the new Parliament building was.

Why didn't the English want the new Parliament building to look new? Because the English hate anything that looks new or clean or modern.

They used to have a surefire way of making things look old: they burned coal for fuel, which quickly left such a thick coat of soot on everything that even new buildings seemed worn after a week or two. People were delighted with this until one day somebody pointed out that burning coal for home fuel maybe wasn't too good for people's health and should be banned. This led to the Great

Debate of the 1950s, when the people of London had to decide if they preferred to cough and wheeze their way through life or if they preferred to breathe. As the British are an eminently sensible people it took them only a few years to decide in favor of breathing. And in 1956 they banned the burning of coal in London.

It was shortly after this that someone had the bright idea, now that the air was no longer sooty, of cleaning up the government's dirty buildings. This started another big row. The Old Guard made the argument that London just wouldn't look like London if everything was clean, which was undoubtedly true, but as everyone knows from reading their history books, groups that are called the "Old Guard" always lose and this time did, too.

Most people were delighted to see what their city really looked like underneath all the soot. But a lot of the oldtimers (that's the way to refer to the Old Guard when you want to show a little sympathy) never reconciled themselves to the change. What they missed most was seeing the city's bright red buses going past dirty sooty buildings. They had warned people the buses wouldn't look half so impressive anymore and they were right.

The problem with Britain, you may be starting to believe, is that nothing is as old as it seems. This is almost true. The pageantry isn't as old, the Houses of Parliament aren't as old. Not even the Old Boy network is really old. Historians say it didn't emerge until the late nineteenth century.

To be sure, the members of the British establishment had always had in common the "old school tie." But nobody had made much of it. Which school you went to meant nothing. Either you were a member of the Establishment or you weren't.

So why did the school you'd attended suddenly become important? In the nineteenth century aristocrats began to worry that with the "rise of the middle classes," it was becoming unclear who was and who wasn't a member of the Establishment. The solution? Thenceforth people who went to name schools were to be considered members of the Establishment and people who didn't were to be considered outsiders. Thus was born the "old school tie."[9]

Much older than the Old Boy network is the House of Commons. But it's not as old as people used to think. Formerly it was believed that the institution went back to the days of the Anglo-Saxon kings, to the fifth or sixth centuries. But it actually had its origins in the eleventh century under Henry II. To me and you this may seem like a long time ago. But to the British the discovery that the Commons went back "only" to the eleventh century came as a great shock.[10]

OF KILTS AND BAGPIPES

Another mistake people make about British history involves Scotland. But before I get to it I want to state firmly my conviction that the history of Scotland is important even if nothing ever happened there. For one thing, it was the birthplace of Sir Walter Scott, whom I personally have never found very interesting, but whom the experts say is really exceptional. For another, if it weren't for Scotland, Great Britain would be known simply as Britain.*

The mistake people make is thinking that Scots-

---

* Historical note: In 1707 Scotland united with England. This gave the British a big head and ever after they went around calling their country Great Britain. This wouldn't have been so bad by itself, but the British have made a bad habit of changing names and confusing everybody. In Roman times Britain was known as Britannia. After the Norman Conquest the name England came into use. In the civil war they called themselves the Commonwealth. With the Restoration they went back to being Britain. After changing to Great Britain, they eventually became the United Kingdom. And people laughed when Burma changed its name to Myanmar.

men always ran around in plaid skirts. They ran around, I'm sure, but they didn't wear skirts (plaid or otherwise).

Scotsmen didn't begin wearing the "traditional" plaid kilt until the eighteenth century. Before that they wore plaid, to be sure, but they did not wear kilts. They wore long knee-length plaid shirts belted in the middle.

It gets even worse. The kilt—the Scotsman's pride and joy—was invented by an Englishman, Thomas Rawlinson, around 1727, near Inverness.

Why did Rawlinson invent the kilt? Why would anyone dream up such a costume? This is the question which has puzzled millions for years and which now can finally be answered. He invented the kilt because the average Scotsman was so poor he couldn't afford a pair of pants.

As it happened, the kilt proved almost instantly to be a great popular success. Maybe it wasn't as good as a pair of pants, but if you were used to wearing a knee-length plaid shirt belted in the middle you'd have run out to get one, too.

Why Rawlinson became interested in the clothes Scotsmen wore is an interesting story in itself. It seems that Rawlinson, an industrialist who'd built an ironworks in Scotland, felt that the lumbermen he'd hired were hindered in their work by the clothes they wore. So he invented the kilt. I know it doesn't make much sense, but neither does a lot of history.

What I can't understand is why Rawlinson didn't just give his workmen a raise so they could afford a pair of pants. I guess it never occurred to him.

The next turning point in the history of the kilt

came in 1745. This was when the British Parliament, successfully proving that there are endless ways for governments to embarrass themselves, decided to ban the kilt. The kilt is in reality just a piece of cloth, but the Parliament had come to believe the kilt was a threat to the British Way of Life. It takes a lot of imagination to understand how people came to believe this, but they believed it. Their objection to the kilt, as far as I can tell, was not that it may have looked a little silly, but that it gave Scotsmen the idea that they were different from Englishmen at a time when the Scottish were in rebellion against the English. Scotsmen were different from Englishmen, but the English didn't want to hear of it. This shouldn't really surprise us. The English spent the entire nineteenth century trying to make people the world over dress, speak, and eat just like them. So I suppose it was just natural that they'd want to ban the kilt.

Prior to 1745 Scots regarded the kilt with little affection. Indeed, because kilts were used mainly by workmen, the members of the upper classes wouldn't ever wear them.

And then there was the ban of 1745. Naturally, as soon as the kilt was banned everyone wanted to wear one. And almost overnight the lowly kilt instantly became the revered national costume of the Scottish people.

Furthermore, as soon as the kilt became a national treasure it was claimed that each of Scotland's chief clans had always been known for a distinctive plaid kilt pattern. The thing hadn't existed the previous century but suddenly people were arguing about which clan had the right to wear which "ancient" pattern.

Surely, you may be wondering, somebody must have stood up and pointed out that the kilt was not an ancient form of Scottish dress and that the clans did not have any claim on one plaid pattern or another. One person did, a scholar named John Pinkerton. Nobody, of course, paid him any mind.

I wish I could say that Sir Walter Scott wasn't taken in by the ruse, but he was. Indeed, Scott himself is responsible in part for the widespread belief in the mythical antiquity of the Scottish kilt. In an essay in 1805 he advanced the claim that it could be traced clear back to the third century.

And as long as we're discussing kilts, what about bagpipes? They, too, I'm afraid, are of recent origin. In ancient Scotland troubadours played the harp, not bagpipes.[11]

PART 7

LET THEM EAT BRIOCHE!

JOAN OF ARC

LOUIS XIV

MARIE ANTOINETTE

ROUSSEAU

VOLTAIRE

LAFAYETTE

NAPOLEON

ALFRED DREYFUS

Marie Antoinette

JOAN OF ARC

In picking Joan of Arc as their national hero the French made the mistake a lot of nations do. They settled on a human being.

Joan of Arc, to be sure, was one of the outstanding representatives of the species. As a teenager she commanded an army and defeated the English in a battle that proved to be a decisive turning point in the Hundred Years' War.

But she could be pretty peculiar, even for a human being. I think, for instance, I am not too far out of line in saying it was peculiar for a teenager to think she could persuade the French dauphin Charles to give her command of an army, even if, as seems to be the case, she believed it was her destiny. (And speaking of peculiar, I would say it was peculiar of Charles to accede to her demand. But I guess he was desperate.)

And I think it was peculiar for her to claim she heard voices. Of course, a lot of people heard voices then.* But she put more stock in her voices than a normal person would. She said the voices told her how to defeat the English.

Some days she heard voices and some days she didn't. It must have been very confusing for her.

Take, for instance, the time she was preparing to march an army on Paris. Just then would have been a perfect time to hear voices, but the voices didn't come. She'd cock an ear heavenward, hoping for guidance in preparing her strategy, and NOTHING. Not a peep.

It was also unusual that she went around dressed in men's clothing. But this I can understand, at least. I suppose she had to disguise her sex, as I don't think teenage girls were in much demand then as soldiers.

The one part of her story that does not strike me as weird is that she ended up being burned at the stake as a witch. It's kind of how you expect the story of a medieval girl soldier who hears voices to finish.

Of course, she shouldn't have been executed. But she was daffy, wasn't she? Say what you will about her trial, the judges seemed to have figured that much out.

Daffy or not, she was a bit of an odd choice as a national hero. She wasn't just fighting the English, she was fighting fellow French, the Burgundians, who ruled Paris and who were at war with the French king.

---

* A lot still do. Some are locked up in rooms with rubber walls. Some can be seen on Sunday morning television.

She wasn't, incidentally, captured by the English. She was captured by the Burgundians and turned over to the English.

The sad truth is, Joan of Arc, the "Savior of France," was done in by her fellow Frenchmen, and Parisians, no less. Like I said, her story is about the most peculiar I ever heard.*[1]

---

* Parisians later tried to make it up to her. They built a huge equestrian statue of her and put it next to the Louvre. Every so often they give it a nice new coat of gold leaf.

LOUIS XIV

No, Louis XIV did not say, "I am the state." He wasn't bright enough to think of it.

Why, if he didn't say it, do people think he did? Voltaire is to blame. He included the bogus quote in his biography of the French monarch. But Voltaire apparently just made it up.[2]

MARIE ANTOINETTE

Marie Antoinette is famous for saying, in response to complaints that the poor were starving, "Let them eat cake." Why is she famous for this? Because she was cold and heartless and it sounded just like her.

Actually, though, she didn't say it. I'm afraid her remarks were never that quotable.

It took scholars a long time to track down the quote but finally they did when one of them, a brave man with a hard ass, actually read all the way through Rousseau's *Confessions,* which is one long book, as Rousseau had a lot to confess, and found that Rousseau had attributed the line to a "young princess." We do not know the name of *this* cold, heartless woman, as Rousseau was a gentleman and did not believe in naming names, except for the names of those he slept with. But it could not possibly have been *our* cold, heartless Marie, as she was not yet born at the time the remark was supposedly made.

Incidentally, the quotation, properly rendered, is "Let them eat brioche," a typically fancy French way of saying "Let them eat cake."[3]

ROUSSEAU

Jean-Jacques Rousseau was a famous Frenchman who loved nature. He had tried loving people, but things just never worked out.

It was always the same with Rousseau. People were delighted to be introduced to him. Then he'd invite himself in as a guest for several years, and before long people didn't think so highly of him anymore. Rousseau, though, never could figure out what went wrong. He was a little dense that way.

One of his biographers says his problem was that he never learned the art of conversation, which made him a dull dinner guest. I think, however, it went a little deeper than that.

His relationship with Diderot was typical. At first, they were great friends. Then, as happened a lot with Rousseau, he and his erstwhile friend couldn't stand each other. The trouble began when Rousseau attacked Diderot's philosophy of rationalism. Rousseau didn't mean for Diderot to take the attack personally, but Diderot, being human and all, couldn't help it.

Rousseau also got along with Voltaire at first.

Then they started saying mean things about each other and that about ended their friendship. Between the two of them, I'd have to say Voltaire got off the meanest attack. He accused Rousseau of having given away his own children. It happened to be true, but nobody likes that kind of thing to get around.

Why had Rousseau given away his children? Because he was heartless? Because he didn't care for his own flesh and blood? Rousseau said he did it for their benefit. He knew he wouldn't make a good father, so he figured they'd be better off without him.

There is not much of a defense for what Rousseau did, but there is one. Apologists have pointed out that it was practically expected in Paris in his day for single parents to give away their children. A recent study revealed that in 1772, a third of the babies born in Paris were abandoned.

Having alienated just about every living person in Paris and London, Rousseau retreated to a remote area of Geneva, where he lived out the rest of his life. It was there that he wrote his great masterpieces, including the *Confessions*, a book he found easy to write, as he had plenty of material to work with. I myself think it's a bit long. I believe he could have told half as much and the world would still have gotten the general idea.

Rousseau is best known, of course, for his celebration of the "noble savage." I think it's worth pointing out, though, that he himself never met one. He got all he knew out of a book he happened upon. This wouldn't have been so bad except that he happened upon the wrong book. It was by

François Coreal, a now-notorious travel liar who contrasted, in Percy Adams's words, the "primitive innocence of the American Indian and the decadent corruption of the European interloper." In Coreal's account, the Indians never do anything wrong, the Europeans nothing right.

About Rousseau's political philosophy there are a number of myths. I'll mention just two.

No, Rousseau was not a democrat. He believed people should be ruled by those who are smart, which, when you think about it, sounds like a good idea.

No, Rousseau was not a free thinker. He believed in God. He did, though, have trouble making up his mind which God to believe in. One year he was definitely sure it was the Calvinists' God he liked. The next year, afraid he'd made the wrong choice, he'd switch to the Roman Catholics' God. I think at the end he felt like he'd prayed to enough different Gods that it was finally safe to move on.[4]

VOLTAIRE

The statement with which Voltaire is most identified—"I disapprove of what you say, but I will defend to the death your right to say it"—is a twentieth-century invention. It was made up by Beatrice Hall (pseudonym: S. G. Tallentyre) in a book published in 1907. Hall never said Voltaire said it, she only said it was something he might have said, but of course that didn't matter. Forever after, Voltaire was cemented to the quote and it to him.

Of late it's been claimed that unbeknownst to Beatrice Hall, Voltaire actually expressed the sentiment she ascribed to him. Ashley Montagu says he heard this from Leo Rosten. Rosten reportedly said he read it in a book by Norbert Guterman. Guterman wrote that he found the quotation in a letter Voltaire wrote on February 6, 1770, to Louis Henri Leriche. But the letter to Leriche does not include the famous saying or anything like it. What it does include is Voltaire's other famous quotation: "God is always on the side of the big battalions."

Oh, and incidentally, Voltaire did not coin the expression "God is always on the side of the big battalions." Voltaire wrote, "*It is said* that God is always on the side of the big battalions."[5]

LAFAYETTE

The Marquis de Lafayette isn't in the same league as some of the others included in this section. But he was French. He was a hero. And there are some myths about him.

Lafayette is famous because he is one of only three prominent Frenchmen ever to visit America who liked the place.*

He is also famous, of course, as the Frenchman who helped George Washington win the war for freedom and liberty and all that is good in the world. But one thing I never did figure out was why he didn't drop the title Marquis. If he was as committed to democracy as he let on, I think he would have gone around as Bob, or something.

The chief myth about him is that he came over here to show his sympathy for the American cause. Actually, there is every reason to believe that he really came over to kill as many British blokes as he could. It seems the British had killed

---

* Another one was de Tocqueville. The third was J. Hector St. John de Crevecoeur, an eighteenth-century farmer.

his father in the Seven Years' War and Lafayette never forgot it.

He also wanted what every man who is young and stupid wants from war: glory. The "one thing for which I thirst," he wrote Washington, "is glory."

That he was a genuine hero, I have no doubt. But I'd feel better about him if he hadn't left his pregnant wife when he came over here.

Lafayette did do a lot of good. But his goal in going to America, as historian Esmond Wright concludes, "was personal independence and self-assertion, not American freedom."[6]

NAPOLEON

The main belief about Napoleon, that he suffered from a Napoleonic complex and wanted to rule the world, is almost always attributed to his short height. But he wasn't short. By contemporary standards, he was of average height. The confusion about his height was due to the fact that it was widely reported, after his autopsy, that he measured just five foot two. But the five-foot-two figure was based on the old French system of measurement, known as *pieds de roi.* Using the modern standard of measurement, he was actually a little over five foot six.

The experts have provided their own explanations for his ambitiousness. Freud said he had a Joseph complex, which was said to arise from a compulsive desire to outdo his brother. Another expert ascribed his ambitiousness to a "churning pituitary" gland. Yet another claimed he had a hyperactive thyroid. Take your pick.

Napoleon's characteristic pose, his hand stuck smartly in his vest, has fascinated people for generations, giving rise to all sorts of interesting psychological diagnoses. All are in error, however. His behavior was actuated by a physical cause, not a psychological one. His whole adult life he suffered from sharp stomach pains. Keeping his hand on his stomach helped relieve the affliction.

A few years ago one of the television networks promoted a miniseries about Napoleon's marriage to Josephine as "the greatest love story ever told." Actually, Napoleon married Josephine for her connections. A second-class Corsican, he needed her help in gaining entry to the circles of the ruling elite. In time he fell in love with her, but when she didn't give him an heir, he divorced her and married someone else. To the end of his life he insisted only she understood him. But he always distrusted her, some might say with reason. When he went away on his campaign in Egypt, for example, she had affairs with at least two men. But Napoleon couldn't complain he'd been cuckolded on the sly. Josephine openly consorted with her young lovers.[7]

Only the French recall the emperor's glory days. The rest of us seem to savor his demise, remembering him for his defeats: the Moscow retreat and Waterloo, both of which are suffused in error.

The cruelty of Russian winters is so well known that the story of Napoleon's defeat in the winter of 1812 is taken almost as a given. The question becomes, in most people's minds, not why he was defeated, but why he ever attempted an invasion in the first place, and having attempted it, why he left so late his army inevitably had to fight through a brutally cold winter.

Actually, the winter of 1812 in Russia was

remarkably mild. The army left Moscow October 19. The first severe frost didn't arrive until October 30. And the temperature didn't drop into the teens until November 12, and then only temporarily. In late November there was a thaw. The reason the famous crossing of the Beresina on November 26 was so deadly was that the stream had melted, trapping French troops on one side until Napoleon could build a makeshift bridge for their escape. Winter temperatures didn't drop below zero until December 4.

We blame Napoleon's defeat on the Russian winter because Napoleon himself did, in order to lessen his own responsibility for the failure. But the army was broken long before the winter cold arrived. Napoleon had left Moscow with nearly 100,000 troops. By November 12—the first day the temperature dropped into the teens—only 41,000 were left. It wasn't winter cold which killed the army, it was disease. Undoubtedly the winter weather had weakened Napoleon's men, but the temperature was hardly cold enough to have killed them. George Washington's army, a few decades earlier, survived far worse weather.

Napoleon undoubtedly would have been better off if he'd left before the onset of the cold weather, if he were set on leaving. But his real mistake wasn't leaving late, it was deciding to leave at all. Biographer Vincent Cronin is of the opinion Napoleon could just as well have stayed put. Moscow was safe, most of the Russians had evacuated, there were plenty of supplies, and in the spring he probably could have scored a decisive victory over the czar. The very quality, however, which made him a great leader—impatience in getting things done—also probably contributed to his great mistake in Russia. He just couldn't bear sitting still.

The oddest finding of historians is that his army probably suffered as much from the heat as the cold. The Russian summer of 1812 was so hot that tens of thousands of his soldiers died from heat exhaustion and sunstroke.

But it wasn't the heat or the cold or disease that ultimately did in the Grand Army. It was, says historian David Chandler, its size. The army—655,000 strong at the outset—was simply too big to lead through a hostile land, making it impossible for Napoleon to feed and supply his force properly.

The story of Waterloo always comes out the same way no matter who tells it, as the decisive defeat of Napoleon's career.* But if everybody agrees Napoleon lost the battle, there's disagreement over who won it. The English and Americans say the Duke of Wellington won it. The Germans say General G. von Blücher** won it. The Belgians say it was their man who won it. The Belgians note that it was only because one of their generals ignored Wellington's order to retreat that the English won. Hence, in their texts, Belgium defeated Napoleon!

---

* Except in one curious account, the one written by Napoleon himself, as A. J. P. Taylor astutely points out. In his memoirs Napoleon persuaded himself that he had won the Battle of Waterloo. His review of the battle ends with an expression of sympathy for the people of London, "when they learnt of the catastrophe which had befallen their army."

** The commander of the Prussian army, who came to Wellington's rescue.

Never before heard of Blücher's role in the battle? Don't fret. The German accounts don't usually mention Wellington's.

The explanation of Napoleon's defeat at Waterloo has long been the subject of controversy. Some have said he was distracted by painful hemorrhoids. Of late it's been claimed he suffered from sleeplessness, owing to (his own) loud snoring. But he was probably in good health. Biographer Vincent Cronin says, "the one surviving order in his own hand is neatly and clearly written: always, with Napoleon, a sign of physical and moral well-being." Why then did he lose? According to Cronin it was because: (1) he spent the critical morning of the battle inspecting the wounded when his army should have been fighting, (2) he underestimated the English, who'd studied his traditional tactics and learned how to neutralize them, and (3) he'd been overconfident, waging war on the assumption the worst wouldn't happen. But the worst did: the Prussians, whom he thought he'd defeated for good two days earlier, succeeded in coming to Wellington's rescue.[8]

ALFRED DREYFUS

Alfred Dreyfus is the famous Jewish French army officer who was framed for treason in the late nineteenth century. None of his story made much sense, but it all actually happened just about as everybody thinks it did. He actually was accused of spying for Germany for money, though he was the scion of a rich French family and had no need for a bribe. It was actually an anti-Semitic investigator who, sensing a cover-up, doggedly pursued the case and proved Dreyfus was innocent. And the military did convict Dreyfus twice, even though by the second trial all knew that the evidence used against him had been forged.

But if Dreyfus was a martyr, as is universally maintained, he was a curious kind of martyr. For while friends and supporters always insisted he'd been framed because he was Jewish—as undoubtedly he had been—Dreyfus himself never agreed. Indeed, he rarely mentioned that he was Jewish.[9]

PART 8

LIKEABLE (AND NOT-SO-LIKE-ABLE) FAMOUS PEOPLE

MACHIAVELLI

CATHERINE THE GREAT

SUN YAT-SEN

CHIANG KAI-SHEK

GANDHI

Devil reading Machiavelli's The Prince

MACHIAVELLI

Niccolò Machiavelli has only himself to blame for the bad things said about him. Go around saying it's necessary for politicians to lie, steal, and cheat and some people are going to get the idea you yourself believe in lying, stealing, and cheating.

He has also had some unfortunate admirers: Cesare Borgia, Mussolini, and people of that ilk. With admirers like that, a man's innocence is apt to be brought into question a little.

And I don't think it was helpful for him to point out that the Borgias weren't all bad, even if he sincerely believed it. The fact is they were bad enough.*

But Machiavelli is misunderstood. Truth is he was a swell fellow. A Republican and a statesman, too. It may be that as a thinker he expressed a high tolerance of political shenanigans, but in his own career in politics he never once resorted to anything of the sort. If there was somebody who, in

* Lucrezia, it is now said, got a bum rap. But Cesare and Rodrigo, I hear, were as bad as they say.

the public interest, needed to be beat around the head he always let the other fellow do it.

The Prince, his masterpiece, is also misunderstood. Machiavelli does not say in there that politicians ought to act immorally. What he says is they sometimes have to.

Besides, Machiavelli is hardly to blame for the fix in which we find ourselves today, with politicians eating us alive and all. I think this would be pretty much the same old crummy world it is whether Machiavelli had written *The Prince* or not.

In short, Machiavelli did not invent Machiavellianism. Nobody is bright enough to have pulled that one off.

Why did he become persuaded that Machiavellianism is okay? Because, of course, he'd started out as a rosy idealist and you know how they always end up.

Machiavelli, incidentally, was a smart man. After he wrote *The Prince,* he hid it in a drawer, where it stayed until after he died.[1]

CATHERINE THE GREAT

Catherine the Great is remembered as one of Russia's most famous rulers, but she wasn't Russian. She was German. And her real name wasn't Catherine. It was Sophia. You'd think maybe all this would have made her people suspicious, but nobody seems to have ever raised any questions.

She came to the throne after throwing her husband, Peter III, in jail. This, too, you might think, would have warranted debate, but it didn't. Peter didn't even complain, but then, he wasn't in a position to. Shortly after his incarceration some of Catherine's friends paid him a visit and found him dead in his cell. Catherine announced he'd died of alcoholism and colic and everybody went away satisfied.*

Some people have made fun of her because she went through so many lovers. But occasionally it redounded to Russia's credit, like the time she fell in love with a Pole and had him installed as the

---

* Peter had picked the wrong person to marry. Like many people, he found out too late.

king of Poland. This didn't do the Poles much good, but it helped Russia immensely. A short while later she asked for the eastern part of Poland and he willingly turned it over to her.

In 1768 the Turks declared war against her, thinking she was a pushover. They later regretted this—when she defeated them, winning the Crimea in the process.

All in all she had a glorious thirty-four-year reign. She even managed to get good notices in Europe as an "enlightened despot."

More than any other ruler of her time she is associated in the public mind with the Enlightenment. She helped rescue Diderot from poverty. She corresponded for years with Voltaire. Several of the *philosophes* dedicated their books to her. She never did manage to get around to helping the serfs, though. In fact, historians unanimously agree the serfs were worse off at the end of her reign than at the beginning. But she meant well.*

If we think of her as enlightened it's largely because Voltaire did, but the historians say Voltaire had been hoodwinked. He'd even believed her claim that the Russian peasants all ate chicken. What Voltaire forgot to ask was: how often.

She did believe in reason and reform. But she wasn't too hot on revolution. To a smart woman like Catherine it didn't seem like something she ought to embrace. She was probably right about that.[2]

---

* Her apologists insist she would have helped the serfs if she could have but she just didn't have the power to do so. The nobles really ran things. Poor helpless Catherine.

SUN YAT-SEN

Sun Yat-sen is the famous revolutionary who brought down the Manchu dynasty in the revolution of 1911. Only it seems, when the matter is gone into, he didn't. He was actually in Denver, Colorado, when the revolution took place. He found out about it at breakfast when he opened his morning paper.

Why, then, is he given credit for starting the revolution of 1911? Because Sun convinced people he had. He was a genius at self-promotion.

His greatest publicity coup came in London some years before the revolution of 1911. It happened when Sun found himself under arrest inside the Chinese legation in London. From inside the legation Sun orchestrated such a powerful newspaper campaign on his own behalf that the Chinese eventually were forced to give him his freedom.

In the hubbub over his release people forgot to ask how Sun happened to have found himself locked up in the Chinese legation. It came about like this: He was walking by the legation one day when he decided to go in. And when he went in, they nabbed him, just like that.

You could argue that maybe he shouldn't have gone into the legation, seeing as how he was a fugitive from Chinese justice and all. But Sun just couldn't help himself.*

In many books it's recorded that Sun Yat-sen had attempted ten coups against the Manchus over the years, beginning in 1895. This is true. But his attempted coups never seemed to have too much chance of success. One failed because his supporters missed the boat—literally. They missed the boat that would have taken them from Hong Kong to Canton. Another failed when his supporters showed up at army headquarters and were slaughtered. They had expected to get help from mutinous soldiers located near the headquarters, but they forgot to tell the soldiers this. That is, they had made elaborate arrangements with the soldiers. But they forgot to mention when they'd be arriving. You can see how this might happen, of course. In a revolution there are a lot of details to be handled. Some are bound to be overlooked.

Sun organized the coups in the name of democracy. But he had a funny habit of relying on somewhat unsavory characters to help carry them out. In the 1895 coup attempt, for instance, his motley crew of rebels consisted of "riffraff, bandits, secret-society strongarms, and demobilized Chinese soldiers." Later he got help from Chinese drug lords made rich through the sale of opium.

Americans are realists, of course. We are happy

* He had a disguise on. But somebody happened to spot him. It was just plain bad luck.

our revolution was made by men of honor. But we accept the fact that other revolutions may not be.

Nobody ever said Sun Yat-sen was George Washington, anyway. But people did compare him to Abe Lincoln. In 1942 the United States government published a five-cent postage stamp with both their pictures on it. The caption read: "OF THE PEOPLE, BY THE PEOPLE, FOR THE PEOPLE." Maybe he wasn't really the Abe Lincoln of his people, but there was a war on, and it was important to try to help the Chinese defeat the Japanese, who'd invaded Manchuria.

Americans always liked Sun. He spoke English and was a practicing Christian, details that help if you are a foreigner and want to be liked by Americans.

Sun returned the love Americans showed him. In fact, he even forged an American birth certificate. Some have said he did it only so he could obtain American protection after one of his coups failed. I'm sure they're wrong about this.

His love life was a mess, but then, so are most people's. What happened to him could happen to anybody. He fell in love with a twenty-year-old girl when he was fifty. His best friend's daughter. A girl he'd been like an uncle to her whole life. And he was still married to his previous wife when he married her. It could happen to anybody.[3]

CHIANG KAI-SHEK

Chiang Kai-shek is remembered as a dictator, but not as a really bad one. He went around killing people, but they were the wrong kind of people, so it didn't really matter.

Like most aimless young people the world over, he started out in life by joining a gang. This was an exciting experience for him, except he got a little more used to murdering and robbing people than was altogether good for his moral development. By the time he was thirty, says Sterling Seagrave, British authorities in Shanghai had indicted Chiang countless times for robbery and extortion, and once for murder.

How many people he actually killed as a gang member, no one knows. Seagrave guesses at least three. Each time, though, he had his reasons. Take the time he killed a patient in the hospital who'd been recuperating from a painful illness. The man had dared to get into an argument with Chiang, so Chiang, naturally, pulled out a gun and killed him. And he had a perfectly good reason for killing a rival gang member. Doing so made it that much easier to frame the fellow for another murder Chiang himself had committed.

Chiang showed so much promise as a gang member that he quickly became the favorite of the leader of the Green Gang, Big Eared Tu, a colorful opium addict with a shaved head. Together they proceeded to conquer China and become rich and powerful.

You don't hear much about Big Eared Tu in the standard biographies of Chiang Kai-shek. Biographers must want readers to believe Chiang rose to the top all by himself, I guess.

But who, for instance, do you think got Chiang his first big break, his appointment as head of Whampoa Military Academy in 1924? Why, it was Big Eared Tu, of course.

Some have expressed their disappointment that Chiang turned out to be corrupt. I think they miss the point. He never could have gotten where he was in life by being honest. It's not like he started out pure and became corrupt. He was always corrupt.

Having a guy like Big Eared Tu on his side proved exceedingly helpful through the years. Like when the bankers in Shanghai threatened to depose Chiang. Who "persuaded" the bankers to keep Chiang in power? Big Eared Tu.

And in the 1930s, when Chiang was in need of airplanes for the government, who reached into his own pockets to buy them? Big Eared Tu. In all, Tu purchased 120 planes.

And when Chiang needed money to pay for the army, who helped him out? Big Eared Tu.

Where Tu got the money to help out his pal Chiang has never been gone into in any great detail, as Tu wasn't the type to file annual reports. But I think it's safe to say he made most of his money selling drugs.

Which reminds me of a little joke Chiang played on the Americans. The Americans had demanded that Chiang put a stop to the opium trade, as it was leaving the Chinese in a bad stupor, and as Americans were no longer making a buck off the darn thing. Chiang said he'd be glad to. Whereupon he promptly established an Opium Suppression Committee and placed Big Eared Tu in charge.

To his credit, Chiang tried to put an end to some of the corruption. At one point he told Big Eared Tu he wouldn't pay Tu any more protection money. Then Tu had Madame Chiang Kai-shek kidnapped. That was the last anybody ever heard of Chiang's drive to end corruption.

I wouldn't blame Big Eared Tu for Chinese corruption, though. I think even if Chiang had eliminated Tu there still would have been corruption, as it seemed to pay pretty nicely. Like when a millionaire businessman in Shanghai paid Chiang a $200,000 ransom to get his son back after the army kidnapped him. Or when another businessman paid $500,000 to get *his* son back. Or when yet another businessman paid $700,000 to get *his* son back.

Speaking of Madame Chiang Kai-shek—as I was a paragraph back—did it ever seem funny that *Time* magazine referred to her as "the Missimo"? I had a hunch you'd feel that way. How about *Time*'s practice of calling Chiang himself "the Gissimo"? If you ask me, that was a pretty wacky way to refer to the leader of a powerful country.

And speaking of wacky, I can't help but think of Madame Chiang's brother-in-law, H. H. Kung, who was one of the richest businessmen in the whole world. I would say it was pretty wacky of him to write Hitler a letter saying that Hitler was "a model for us all, a great fighter for rectitude, national freedom and honor."[4]

GANDHI

Gandhi used to go around saying, "I am a true mahatma."* And I'm sure he was. But even true mahatmas can behave strangely.

One of the strangest things about him is that as an old man he liked to sleep in the nude with naked young women. Maybe, on second thought, this wasn't so strange.

Gandhi explained that he did this to test his vow of chastity. If he got any pleasure out of it at all he never admitted it.

He had a wife and he could have slept with her in the nude, but sleeping with her, apparently, wouldn't have been much of a test. He didn't seem to be taken with her looks. "I simply cannot bear to look" at her face, he once remarked. "The expression is often like that on the face of a meek cow."

Gandhi wasn't against sex, not at all. It was his belief that a married couple should have sex three or four times—in one lifetime. He said there

* Mahatma means "great soul."

should be a law against couples having sex more than that.

He himself had enjoyed lots of sex. But that was when he was young. He didn't want anybody else to make the same terrible mistake he had.

His opinion on ejaculation was that it should be avoided: "Ability to retain and assimilate the vital liquid is a matter of long training. When properly conserved it is transmuted into matchless energy and strength." When he awoke one morning and discovered he'd accidentally had a nocturnal emission he is said to have almost suffered a nervous breakdown.

He was also concerned with bowel movements. According to one biographer, they are the subject of much of his correspondence. What intrigued him especially was the usefulness of enemas. He himself had one every day. He had his young women take a daily enema, too. (His daily greeting to them was: "Have you had a good bowel movement this morning, sisters?")

His view on bathrooms, as you might expect, was firm. They must be absolutely clean! "The bathroom is a Temple," he wrote. "It should be so clean and inviting that anyone would enjoy eating there."

Like a lot of great men Gandhi treated his family badly. As you know, he wasn't too wild about his wife, so maybe it's not a surprise he never let her get an education. But he never let his sons get an education either. And he disapproved of his eldest son's marriage and disowned him.

He hated modern medicine. He hated it so

much he wouldn't even let British doctors administer a life-saving shot of penicillin to his wife when she came down with pneumonia. It was a tough choice he faced, saving his wife or sticking with his principles, but Gandhi could be tough when he had to be. His wife died, but he still had his principles.

I don't want to leave the impression he was a fanatical opponent of modern medicine. Sometimes even he could see its uses, like the time, just after his wife died, when he let doctors give him quinine to help him get over a bout with malaria. Or the time he suffered an attack of appendicitis and agreed to let the surgeons perform an appendectomy on him.

His reputation for peaceableness is well deserved, but it came a little later in life than most people suppose. They didn't mention it in Sir Richard Attenborough's movie, but through middle age Gandhi rather liked war. In all he volunteered to serve in three imperial wars: the Boer War, the war against the Zulus, and World War I. He never did get the chance to serve in World War I, though. Just when he went to sign up he came down with a bad case of pleurisy and couldn't join. As it happened, it was a good thing he couldn't. It saved him from having to make a lot of silly explanations two years later when he announced to the world his opinion that the British Empire was one of the great forces of Satanism.

He never was the pure pacifist he's made out to be anyway. He always approved of the use of violence as a last resort. Like during one of the periods when Hindus and Muslims came to deadly blows. You didn't hear this in the movie, but when the Nawab of Maler Kotla issued an order to shoot ten Muslims for every Hindu who was killed in the state, Gandhi gave it his blessing.*

Generally, of course, he recommended nonviolence. When Hitler overran Czechoslovakia, Gandhi told the Czechs it would be better to commit collective suicide than fight. Later, he gave the same sage advice to the Jews.

People tried to convince him the Nazis weren't like the British, that nonviolence wouldn't work with them, but Gandhi always thought it would. One day, after Hitler had already conquered Poland, France and most of the rest of Europe, Gandhi decided what was needed was an appeal to Hitler's conscience. So he sat down, thought hard about how to appeal to a conscience nobody else thought existed, and finally came up with his answer: an open letter. "Hitler is not a bad man," he told friends, "he'll listen." But he didn't.[5]

---

* Along these lines is the statement Gandhi made about independence: "I would not flinch from sacrificing a million lives for India's liberty."

PART 9

KING ARTHUR AND SUCH

KING ARTHUR

LADY GODIVA

ROBIN HOOD

PIED PIPER

WILLIAM TELL

DRACULA

FRANKENSTEIN

LITTLE DUTCH BOY

SANTA CLAUS

William Tell with Lady Godiva

KING ARTHUR

To get to the bad news first: King Arthur had rotten teeth. This is the reason you never see pictures of him smiling. Come to think of it, you never see pictures of anybody from the old days smiling. This is because they all had rotten teeth. They also smelled bad and bit their nails. They never tell you these kinds of things in normal history books because historians are trained to overlook the unpleasant fact that history is about human beings. But this isn't a normal history book.

Actually, I don't really know if King Arthur had bad teeth or not. In fact, I don't know a thing about him. Not when he was born, not where he lived, not whom he married, not when he died or where he's buried. If I knew his shoe size that would be something, but I don't even know that.

And neither does anybody else. Nobody knows

one single solitary fact about him. The reason for this is that Arthur never lived.*

Ever visited South Cadbury Castle in southwest England? They claim that King Arthur lived there, but you know what? They're full of it. They made the whole story up so they could attract the tourists.

Ever visited Glastonbury Abbey, near Somerset? They say King Arthur is buried there. But they're full of it, too.

To be sure, the claim that he's buried there goes back a long ways. Monks in the twelfth century professed to have discovered his remains in a hollow log buried sixteen feet underground. But the monks were lying. They had no idea whose remains they'd found. They may have even planted the remains themselves.

So why did they lie? Apparently it was a publicity stunt. Their monastery had burned down and they needed to attract attention so they could get some help.

Anyway, that's what some people say. Others blame the deception on King Edward I. Why would Edward want it known that Arthur's grave had been found? Because he wanted everyone to know King Arthur was dead. Edward apparently lived in fear that some day somebody was going to show up claiming to be King Arthur. (King Arthur, you see, was supposedly immortal.) And, of course,

---

* "The fact of the matter is that there is no historical evidence about Arthur; we must reject him from our histories and, above all, from the titles of our books." Professor D. N. Dumville, as quoted in Richard Schlatter's *Recent Views on British History* (1984), p. 11.

if King Arthur appeared, people would want *him* in charge instead of Edward.

If we don't know anything about King Arthur—if there's no evidence he lived—why do so many believe he lived? Because somebody led the British defense against the invading Saxons, Angles, and Jutes in the fifth century, and it's possible he was named Arthur and that his fame reached down through the centuries in oral British tradition. But there's not a bit of proof.

Who started the Arthurian madness? It would appear to be the work of an eighth-century Welshman named Nennius. So the story of the greatest English hero there ever was was invented by a Welshman.

Nennius wasn't a liar. He just enjoyed telling good stories. One of his favorites was about the day King Arthur singlehandedly slew 960 enemy Saxon soldiers. It was such a good story the English believed every word of it. Not one Englishman, as far as we know, ever said, "Hey. Wait a minute. 960? That's a lot of dead soldiers."

It wasn't Nennius, though, who turned Arthur into a popular hero. It was Geoffrey of Monmouth, who wrote in the twelfth century.

The reason we think of Arthur as a shining knight surrounded by sweet damsels in distress is because Geoffrey himself, as I mentioned, lived in the twelfth century. And in the twelfth century you couldn't go to the corner store to buy a bottle of milk without running into a knight who'd just rescued some sweet damsel.

Why is the date important? Because Arthur is supposed to have lived in the fifth century, not in the twelfth. And in the fifth century there were no knights. Knights didn't appear in Europe until the eighth century, and they didn't appear in England for several centuries after that. And if there were no knights there was no chivalry, and if there was no chivalry, there were no damsels in distress for Arthur to rescue. There may have been a Round Table, which has been traced back to the Celts, but of course there wouldn't have been any knights seated around it.

Nor, for that matter, did people live in castles back in the fifth century. Stone castles did not appear in England until after the Norman Conquest six hundred years later. So Arthur couldn't have lived in a stone castle even if he had really existed.

Anyway, the Arthur we are familiar with is not the Arthur known in medieval England. Between then and now Arthur's story got sanitized. In the medieval versions, for instance, he commits incest with his half-sister, by whom he had a bastard son.

The story changed in the Victorian era. Victorians, being Victorians, they left the sex out. This had two harmful effects. One, it made King Arthur a little boring. Two, it ruined the story line. In the medieval version his mother and father meet at a dinner party. They immediately fall in love. They have sex. And then she goes home to her husband. In the Victorians' accounts, they skip over this part.[1]

LADY GODIVA

Lady Godiva, if the story told about her is true, was a disturbed young woman in need of serious psychiatric care. Nobody goes about naked on a horse just to lower people's taxes. Somebody who behaves like that has more than taxes on their mind. And I don't care if her golden yellow mane came down just so far or not.

I would also guess that her husband, Leofric, Earl of Mercia, had a loose screw or two as well. I think you have to be pretty far gone to make the kind of deal he is said to have made. What husband in his right mind would tell his nagging wife: "Okay, look, you care so damn much about the taxes the poor people are paying. I tell you what. I'll lower their taxes just like you want if you go romping through town at high noon stark naked." What kind of deal is that?

Actually, though, neither Lady Godiva nor Leofric were crazy. He never told her he'd lower people's taxes if she rode naked through town and she never did it. It's just a nice little story somebody cooked up long after both had been dead and buried.

The only part of the story that's true is that Leofric was rich, Lady Godiva was his wife, and the couple lived near Coventry. End of story. Nobody even knows if Leofric raised or lowered taxes. It's considered unlikely that he would have raised people's taxes considerably. The people were too poor to have been able to pay them.

What about Peeping Tom? He was an afterthought. Somebody dreamed him up in the seventeenth century.[2]

ROBIN HOOD

There have been quite a few people in English history who stole from the rich and gave to the poor (though stealing from the poor and giving to the rich seems to have been far more common), but none, as far as we know, went by the name Robin Hood.

Historians have found a half dozen or so medieval men by the name of *Robert* Hood who pretty much fit the bill. One, who lived back in the 1200s, even seems to have gone around stealing from the rich "for the benefit of the many." But nobody knows if the Robin Hood saga is based on the adventures of this particular fellow or not.

The Robin Hood saga, I should point out, has changed over the years.

That Robin Hood was an aristocrat, for example, seems to have been added into the story in the sixteenth century. So was the business about his love of Maid Marion and his friendship with Friar Tuck. In the original versions of the story, he didn't even roam around Sherwood Forest. He lived in Barnesdale Forest. Why suddenly somebody decided to have him in Sherwood Forest I couldn't tell you.

In some early accounts, Robin is not so much the class-conscious hero of the poor as the tribal leader of the downtrodden Saxons, many of whom remained unreconciled to the conquest of the island by the conquering Normans. Robin, however, was a funny name for a Saxon leader, as it is French.

And just to show you how his story keeps changing, in the medieval versions he wasn't even lovable. Say a fellow he didn't like happened into the forest and by accident crossed his path. Think Robin would just search the man's pockets for gold and stuff and let it go at that? Robin was a two-fisted troublemaker of the first order. At the least he'd box the guy's ears and yank off one of his limbs or something.

What he'd do to clergymen you wouldn't want to hear. I read it wasn't too pleasant.[3]

PIED PIPER

One reason for liking the "Dark Ages" is that people back then came up with the darnedest stories. Like the one about the Pied Piper: A man with a magic flute rids the town of Hamelin of its rats and then, when he doesn't get paid, makes off with 130 children.

I am afraid, though, there was something to the story. First of all, towns in the Middle Ages were often so beset with rats they did hire rat-catchers.

Second, it is well established that more than a hundred children did up and leave Hamelin in the 1200s, as they up and left lots of little towns in the thirteenth century all over Europe, some to join the Children's Crusade, some to help with the founding of new settlements.

So, upon reflection, it's not as crazy a story as it seems.

Why the children of Hamelin specifically up and left is unknown. But whether the children were sent away on the Children's Crusade or dispatched to new settlements, the parents probably felt some guilt. So they blamed the children's departure on the Pied Piper.

We don't know when the Pied Piper story first appeared. But we know that by 1300 it was well established, which means it was likely invented by the very people who let the children go.

The detail about the Pied Piper being a rat-catcher, however, was added several hundred years later. In the original story he was just a chap with a magic flute.[4]

WILLIAM TELL

William Tell, hero of the late thirteenth century, did not exist. He did not shoot an apple off his son's head. He did not help the Swiss establish their independence from Austria. And he did not insult an Austrian official named Gessler, as no Austrian official ever went by that name.

The whole story, every last riveting detail, is bogus, the creation of an inventive (and anonymous) Swiss patriot in the late fifteenth century.

Why do the Swiss believe it's all true? Why did they build a chapel over the spot where Tell supposedly had lived? Why for centuries did Swiss citizens make an annual pilgrimage to the place where Tell allegedly evaded his Austrian captors?

Because they wanted to.

It's like it was with the Americans and Plymouth Rock. The Pilgrims didn't land on Plymouth Rock. The story's apocryphal. But when you're building a new nation and you start getting a little giddy, as Americans did in the nineteenth century when the Plymouth Rock story took hold, you tend to start making up stuff about yourselves. For some reason, it seems to help.

To this day the Swiss take the story seriously. Professor Alan Dundes, a folklorist at the University of California at Berkeley, reports that a visiting scholar who expressed doubts about the historicity of William Tell was actually threatened with death.[5]

DRACULA

Dracula was for real. Bram Stoker didn't just make up the whole thing.

To be sure, Dracula—the real one—did not sleep in a box in the dark during daylight. Nor did he revive himself by sucking the blood of innocent virgins. But he did live in a castle in Transylvania. He looked like the devil incarnate. And he liked killing people.

People always get worked up about vampires, but it is human beings like Dracula who make better killers. The high point in Dracula's career came in the early 1460s when he is said to have killed 24,000 Turks. Dracula undoubtedly felt it was all in a good cause. So may his subjects. Then as now, as historian Craig Conant points out, there wasn't "any love lost between the Christian and Moslem communities of the Balkans."

Folks called him Vlad the Impaler. In an old woodcut drawing Vlad is shown seated next to a human fence made out of corpses impaled on sharp wooden poles—a graphic presentation of how he got his charming nickname.

In his day, as you can imagine, he was the talk of the town. My favorite Vlad story is about the time a couple of visiting ambassadors refused to take off their hats to him, as it wasn't their custom to remove their hats for anybody. Vlad, showing the sense of humor for which he was famous, had their hats nailed to their heads.

His real name, in case you want to look him up, was Vlad Ţepeş. Dracula was just the name his pals used.[6]

FRANKENSTEIN

Frankenstein, I know, they don't include in normal history books. But as I have repeatedly made clear, *this is not a normal history book.*

First of all, Frankenstein is not the name of the monster. It's the name of the mad scientist.

Second, in Mary Shelley's original version, the monster is not dumb. He is the pointy-headed intellectual type: speaks French, reads Milton for kicks, and in his spare time studies history, including Plutarch.

Of course, like any good monster, he whacks a couple of people to death, including the scientist who made him, and other folks who get in his way. But why does he behave badly? Why is he antisocial? It is his parents' fault. They didn't give him the love he needed as a child.

Okay, he didn't really have parents. But the gist of the story is he needed love and affection and when he didn't get it, well, we all know what happens when people are deprived of motherly love. They don't usually turn out too well.

Which brings me to the true bad guy in the

story: the "good" doctor, the monster's maker, who, in the crunch, cuts and runs, leaving the monster homeless for six long dreadful years.

During these years the monster meets some of the local townsfolk. In the movies the townsfolk always come off badly. But in Shelley's account, they behave with intelligence and decency. Why are the townsfolk portrayed differently? Because in the movies the story is set in the Dark Ages when people are supposed to have been stupid and ignorant and dirty. In Shelley's account the story takes place in the supposedly more enlightened 1700s.

In the end, the monster dies, as all monsters must, except where the producer thinks there's a chance for a sequel. But where does he go and die? To the North Pole.[7]

LITTLE DUTCH BOY

Why did the little Dutch boy stick his finger in the dike? It was not because he wanted to save the little town of Haarlem from destruction. It was because the writer who invented him couldn't think of any other way to show her little hero in a good light.

And who invented him? It wasn't, oddly enough, a Dutchman, but an American, Mary Mapes Dodge. Dodge wrote about him in her classic, *Hans Brinker, or the Silver Skates,* which was published in 1865.

Many Dutch people apparently are not even familiar with the story. On a trip to Amsterdam in 1992 I asked half a dozen local residents what they thought of the little Dutch boy who stuck his finger in the dike. Not one at first knew what I was talking about. Upon further questioning only one registered any recognition of the story.

In 1950 the Dutch built a statue of the boy near the locks. It was to mollify the tourists, who all wanted to know where the Dutch boy performed his miraculous feat.[8]

SANTA CLAUS

Good ole Saint Nick isn't a saint. In 1969 the Catholic Church unsainted him.

He has, to be sure, turned out to be a nice old man. But in old Europe he wasn't. There he turned bad children over to the devil. Didn't even give out presents.

Nor did he look the way he does now. He didn't wear a red suit or sport a white beard. All that came later, after he came to America and got a make-over. In Europe he usually went around clean-shaven in drab old clothes.

Nor did he drive a team of reindeer or charge down chimneys. That came later, too.

Nor, for that matter, was he universally well liked. The Puritans considered him a heretical Catholic icon.

I suppose you probably want to know if Santa Claus ever existed. After much research I am happy to report that people who have done real research into this question believe the answer is: maybe. The likeliest candidate is said to be a saint named Nicholas who founded a Christian cult in the fourth century. But as that was a long time ago nobody knows for sure who this guy was.

That Saint Nick was imported to America by the Dutch is believed by almost everybody, including the author of *Legends, Lies & Cherished Myths of American History.**

Actually, Saint Nick didn't arrive in America until the Revolution. The Dutch in New Amsterdam never mentioned him.

So where did we all get the idea that the Dutch were responsible? From Clement Moore, author of "'Twas the Night Before Christmas." And where did Moore get the idea? From Washington Irving. And Irving? He just made it all up.[9]

* Page 168, paperback edition.

PART 10

RELIGION

THE BIBLE

JUDAISM

CHRIST

CHRISTIANITY

Adam and Eve

THE BIBLE

The Bible's been around so long and is considered so sacred one would think by now the book's main stories—about creation, Adam and Eve, Noah's Ark and the like—would be well known. They aren't.

Take Genesis. Everybody "knows" Genesis clearly states when man was created. But it doesn't. In one place it says man was created at the beginning of the week, in another, at the end, on the sixth day.

The reason for the discrepancy, say biblical scholars, is that Genesis was written by different people at different times. Nobody knows why the authors didn't reconcile their accounts. But the scholars say ancient writers didn't seem to worry if the Bible was inconsistent. We do, but they didn't.[1]

Or take the story of Adam and Eve. A mistake in translation, made by Saint Jerome in the fifth century, is responsible for the belief that they were expelled from the Garden of Eden for eating an apple. It was actually for eating "fruit." It's possible the fruit was an apple, but more likely it was a fig.

After all, they used fig leaves to hide their private parts.

The story of Noah and the ark is also misconstrued. In the movies they always show Noah leading the animals into the ark by pairs, each pair representing a different kind of animal: a male and female giraffe, followed by a male and female cow, etc. They never show him bringing in, say, seven pairs of the same kind of animal. But that is precisely what he was commanded to do in certain cases. While he was to bring in a single pair of each of the so-called unclean animals, he was to bring in seven pairs of each of the clean animals. As it says in Genesis: "Of every clean beast thou shalt take to thee seven and seven, the male and his female."*

The evilness of Sodom is well established in the Bible, but the widely held belief that Sodomites had committed homosexual sodomy is pure supposition. In the places where the Bible refers to the sins of the Sodomites mention is made of pride, greed, and wealth, but not homosexuality. "Not in a single instance," says Yale historian John Boswell, "is the sin of the Sodomites specified as homosexuality." True, after Lot brought the angels into his house, a crowd of men demanded that he "bring them out

---

* In ancient times a clean animal was considered one which chewed its cud and was cloven-footed. Where they got this idea from is anybody's guess. But they may have simply decided that clean animals were the ones they'd been accustomed to sacrificing to the gods. Maybe it was just chance, but the animals they traditionally sacrificed chewed their cud and were cloven-footed.

unto us, that we may know them." But there is no reason to infer that the crowd wanted to "know them" carnally. Before the Christian era Sodom and homosexuality were never even associated.

It may well be that the crime of the Sodomites wasn't sexual in nature at all. Indeed, most scholars today think the Sodomites' offense was their inhospitable treatment of strangers. Admittedly, mere inhospitality hardly strikes the modern reader as much of a crime, let alone a crime punishable by the destruction of an entire city. But in ancient times travelers feared for their lives when visiting foreign places, making inhospitality literally a matter of life and death.[2]

Lot's wife may have been turned to salt as a penalty for looking back at Sodom, but maybe not. It's likely the Bible writer meant us to believe she'd been made barren. In Biblical times anybody who was barren was said to have been turned to salt. The explanation for the expression is that salty regions were regarded as barren because "nothing could grow there."

The story of Jonah and the whale is mistaken in two ways. The Bible doesn't say Jonah was swallowed by a whale, it says he was swallowed by a "great fish." And anyway, the biblical writer probably didn't mean Jonah was literally swallowed by a fish. In the Aramaic language (a Semitic tongue related to ancient Hebrew, still spoken in Syria) in which the Bible was originally written, anyone caught in a difficult situation was said to have been swallowed by a great fish. And if anyone was ever in a tight spot it was Jonah, who, against his will, was made to go to the evil city of Nineveh and preach moral reform.*

Some of the difficulty in understanding the Bible is that the Bible itself is inconsistent and contrived.

Take the Book of Kings. The writer of Kings had trouble with historical dates. So much trouble that one historian has referred to the conflicting dates supplied for royal reigns as "the mysterious numbers of the Hebrew kings." Probably the main reason for the mistakes is that the author simply didn't have any reliable records to work from. So he guessed. Example: The Bible says there were exactly 480 years between the Exodus and the building of the first Temple, and exactly 480 years from the building of the first Temple to the building of the second. But as historian Robin Lane Fox has observed: "These totals are almost certainly too neat to be a coincidence: somebody has fiddled the lengths so that the two totals coincide."

Then again, the authors of the Bible didn't share our own concern with numerical accuracy. It's one reason they found it so easy to pass along the miraculous claims that Adam lived over 900 years, Noah over 500 years.

Nor were they concerned with consistency. Consider the Books of Chronicles. In Chronicles it is said that King David had always kept God's

---

* Village atheists, says Bergen Evans, used to make the claim that the Bible couldn't possibly be true because whales have small throats, too small to swallow a whole man. But the joke was on the atheists. Evans assures us that "many whales have throats quite large enough to swallow a man, whether he be prophet, priest, or profane."

statutes and His commandments. By the Bible's own record, this is palpably untrue. For earlier in the Bible it's revealed that King David (in Lane Fox's words) "had coveted another man's wife, seduced her, lied and arranged her husband's murder; his dying words included specific orders to pay off old debts by killing two legacies from his reign and bringing 'his grey head down with blood to Sheol.'" The author of Chronicles is considered so untrustworthy that Lane Fox refers to him as "this splendid liar." Another historian writes that the Chronicler deliberately distorted facts and "invented chapter after chapter with the greatest freedom."

The Book of Daniel (which, incidentally, wasn't written by Daniel) is said to have foretold future events. But as H. R. Trevor-Roper points out, it was actually written "after the events that it pretended to prophesy." This neat trick was accomplished by backdating the book 450 years.[3]

In 1956 a German journalist, Werner Keller, titled his worldwide bestseller *The Bible Is Indeed Correct.** But which Bible? One of the reasons there are so many biblical errors is that there are a bewildering number of versions of Bibles to choose from. There's the original Aramaic Bible, the Greek Bible, and the Hebrew Bible. When the Dead Sea Scrolls have finally been made fully available, we will have the Dead Sea Scrolls Bible. Further complicating the matter is the fact that there were sev-

---

* The book is now sold under the title *The Bible as History.* Over ten million copies have been sold worldwide in twenty-four languages.

eral different Hebrew texts of the Bible and they disagreed with each other. Editors long ago tried to sort out the differences. They didn't succeed too well. "What we now read in the Bible," says Lane Fox, "is the result of padding and reinterpretation."

Moses is said to have written the first five books of the Bible, the Pentateuch. This is yet another error. But we don't know who did. The authorship of the Old Testament is lost to history. Nor do we know exactly when it was written. But it is estimated that the Old Testament was written over the course of at least seven centuries. The second half of Isaiah, for example, was apparently written more than a century after the first.

Like the Old Testament, the New Testament is also riddled with inconsistencies and mistakes. Take the four Gospels of Matthew, Mark, Luke and John. They disagree about the day the Last Supper was held (on Passover or the day before?). They provide conflicting lists of the names of the twelve apostles. They disagree on the place where Jesus came back from the dead (Galilee or Jerusalem?). And, incredibly, they disagree on one of the most fundamental questions of Christianity, how Jesus ended up dead on the cross. In some places it's implied that Jesus was condemned at a Jewish trial, in other places at a Roman trial. Possibly he had two trials. Possibly he had no trials. Mark indicates Jesus had a Jewish trial but not a Roman trial. John indicates Jesus had a Roman trial but not a Jewish trial.

But we don't know, which is why, after the elapse of two thousand years of history, the old argument is still heard that the Jews were responsible for

Jesus's execution. A Roman (Pontius Pilate) condemned Jesus to death. But we don't know if he did it for his own reasons or because "the Jews put him up to it."

It's been speculated that the early Christians blamed the Jews for Jesus's death because they couldn't, given political conditions in the second century, blame the Romans. This argument makes a lot of sense. It wouldn't have done the Christians much good to blame the Romans for the death of Jesus at a time when Christians were trying to reach an accommodation with them.[4]

The most remarkable thing of all may not be that the four Gospels conflict but that people for so long thought they didn't. As late as 1899 Theodor Mommsen, "the great Roman historian," maintained that the biblical accounts of Jesus's arrest and conviction agree together and conform "on their essential points." Not until 1935 did a historian first examine the Gospels in enough detail to show that each told a different story, each story "shaped by its author's own coherent line of presentation."

JUDAISM

About the early history of the Jews we have little information, except what's in the Bible. Thus, much of the conventional story of the Jews is considered unreliable.

An example of this is the story of Abraham. It's possible he existed. But he may not have. We know nothing about him but what the Bible tells us.

Maybe Moses lived, but there's no evidence outside the Bible for him either. To be sure, a nun in the fourth century A.D. claimed she'd seen the Burning Bush, and that "it was still sending out shoots." But there hasn't been a single sighting since.

The story of the Jews' special covenant with God is pure moonshine. It was dropped into the biblical history of Moses long after the rest of the material about him was written. It's been dismissively referred to by experts as "padding."

A linguistic mistake is responsible for the claim that when Moses struck a rock water gushed forth. In Aramaic when someone strikes a rock it's like saying someone's struck oil. All it means is that someone's found something. It doesn't mean they literally banged on the ground and water magically appeared.

Another linguistic error is responsible for the idea, common in the Middle Ages, that Moses had horns. The Bible says that when Moses descended from Mount Sinai "his face shone." But for many years the Hebrew word for shone, qaran, was confused with the word for horns, qeren.[5]

Nor is there any evidence outside the Bible for David or Solomon. The Bible, incidentally, doesn't mention anything about Solomon's mines, though an inspired archaeologist claimed to have found them. But it's likely, if he indeed lived, that he was rich. Records show that other leaders from the same area and around the same time were able to accumulate huge stocks of mined gold.

The distinguishing characteristic of Judaism in ancient times was supposed to have been its monotheism. But in the sixth century B.C., seven centuries after Moses led the Jews out of Egypt, Jews were still said to be practicing polytheism. Historians don't have any direct proof of it but insist "the likelihood is that many, as before, worshipped other gods beside their Number One."

The claim that the Hebrews invented monotheism is uncertain. Both the ancient Minoans and Egyptians may have practiced monotheism first: the Minoans as early as 3000 B.C.; the Egyptians in 1300 B.C. The Egyptian leader Ikhnaton* is said to have converted Egypt to the monotheistic worship

---

* Or Akhenaton, if you use the old spelling. Or Amenhotep IV, if you use the really old spelling.

of the sun god Aton. This was several generations before the Hebrews converted to monotheism. But whether it's accurate to equate the Hebrews' monotheistic worship of Yahweh with the Egyptians' worship of Aton is in dispute. Anyway, the Egyptians weren't as excited about the idea as Ikhnaton. As soon as he died they went back to polytheism. Possibly this was because they didn't understand the concept. Or possibly it was because Egypt, under Aton, hadn't done too well. Under Aton, Egypt had lost a lot of valuable territory to its enemies, the Hittites. Naturally, people felt this was Aton's fault.*

Judaism, indeed, was a long time evolving. As late as the tenth century B.C., several centuries after the Hebrews arrived in Canaan (Palestine), they continued to make animal sacrifices. Not until some seventy years after Christ's birth was Rabbinical Judaism established.

And the Jewish star originally didn't have six points. In ancient times it had five.

A common belief, held by both Jews and non-Jews alike, is that the old Jewish ban on pork was instituted for health reasons. Actually, the objection wasn't that pork could be bad for you, but that food should only be made out of one class of animals, for consistency's sake. And the class of animals which seemed the most logical source of food was the class used for sacrifices. And pigs, by tradition, were not in this class.

---

* It was Tutankhamen (King Tut), Ikhnaton's successor, who restored polytheism.

CHRIST

Jesus Christ is thought to be well known, but isn't. Scholars regularly churn out books which depict him in wildly different ways, from political rebel to ancient magician to "Hellenistic gadfly." Much of what we think we know about him is as imaginary as his innumerable portraits.

We don't even know when he was born. Jesus himself apparently never said when it was and nobody bothered to ask him. By the time people decided this was a fact worth knowing—around A.D. 75, some thirty years after his death, when the Gospels began to be written—it was too late to establish with any certainty. So people guessed.

The scholarly consensus, if you're interested, is that the writers of the Gospels guessed wrong. Scholars put Christ's birth at anywhere between 6 B.C. and 4 B.C.

His birthday is celebrated on December 25, not because there's any evidence he was born then, but because that was the day Roman pagans celebrated the birth of the Persian sun god, Mithra. There are a lot of disputes about Christ's life but on this subject nearly everybody agrees.

The celebration of his birth on December 25 stretches way back, but not as far back as people might think. It wasn't until the fourth century that Christians began honoring the day as Christ's birthday. And that was only in the West. In the East, Greek Christians celebrated January 6, the date of yet another pagan holiday.*

It *is* known where Christ was born, in Bethlehem. Or so the Christmas carols say. Actually, he almost certainly was born in Nazareth.

The misleading impression that he was born in Bethlehem was left by the Gospels. Why? Because it helped persuade people that Christ was the messiah. King David had been born there and it was generally believed that the messiah would be as well. (Micah 5:2: "But thou, Bethlehem, though thou be little among the thousands of Judah, yet out of thee shall he come forth unto me that is to be ruler in Israel.")

The Gospels don't agree on how Christ came to be born there. Matthew says he was born there because his parents lived there. Luke says he was born there because his parents, though they lived in Nazareth, had journeyed to Bethlehem to take part in the Roman census. (We now know there was no Roman census then.)

That the baby Jesus was cradled in a manger (an old word for a feeding trough), as he is in every

---

* Christmas, it's worth pointing out, as it's celebrated today, is largely a Victorian invention. The practice of sending Christmas cards, buying presents, and waiting for Santa Claus's arrival down the chimney, all this was the Victorians' idea.

church play, is traceable to Luke, who may or may not have been accurate. But the other elements in the traditional scene—the stable, the farm animals—these have no foundation in fact or in the Bible. It's possible they were added because it makes sense that if Jesus was born in a manger the manger was in a stable and there were animals present. But historians believe it's more likely that Jesus is featured in a stable filled with farm animals because the Old Testament says the messiah will be born in a stable with farm animals.* Besides, there doesn't seem to be any reason for supposing Luke's reference to the manger was meant to be taken literally. It was probably symbolic. He gave Christ a humble birth to make Christ appear humble.

Whether he was born in a stable or not, the "wise men" who subsequently came to visit would not be regarded as wise today. They were astrologers. Whether there were three of them or not is unknown. All the Bible says is they brought three presents (gold, frankincense, and myrrh). St. Augustine was of the opinion there were twelve.

Whether they were guided by a star or not is in dispute. Matthew says they were, but he may have said so because the Old Testament predicted that the arrival of the messiah would be heralded by a star.**

---

* Isaiah 1:3 refers to the presence of an ox and an ass at "their master's crib."

** Numbers 24:17: "A star shall come forth out of Jacob, and a sceptre shall rise out of Israel." (The Twelve Tribes of Israel are descended from Jacob, one of the biblical patriarchs, whose name was changed to Israel after he wrestled with an angel.)

If Jesus's mother was a virgin when she had him, people at the time didn't seem to be terribly impressed. Nobody mentioned it in the early stories about Christ. Not even Paul, who surely would have welcomed evidence of a supernatural birth. The first reference to the virgin birth was in Matthew, which wasn't written until long after Jesus had died.

Where would Matthew have gotten such an idea? It's just a guess, but it may have been from the Old Testament, where he seems to have gotten a lot of his ideas about the messiah. Once again he seemed to be bolstering his case about Jesus by having Jesus fulfill an Old Testament prophesy. In this case, however, there is a problem. The Old Testament did not prophesy a virgin birth. It prophesied the birth of a son by a "young girl." Matthew, unfortunately, relied on a faulty Greek translation in which the son was to be born to a "virgin."*

And what of his ministry? What did he teach? He did not teach that sex between consenting adults is wrong. He preached that adulterous sex is wrong.**

Nor did he preach against homosexuality. He never seems to have mentioned the subject.

Christ may have believed that "he that is without sin among you, let him first cast a stone," but he probably didn't say it. Historians believe the quotation was inserted into the Bible by an editor

---

* Isaiah 7:14: "A young girl shall conceive and bear a son."

** That, at any rate, is what's claimed by historian John Boswell. Boswell concedes there may be an exception in Matthew 15:19 and Mark 7:21, but in his opinion, in these instances Jesus is condemning extramarital sexuality.

long after the Gospels were originally composed.

Christ's age during his ministry is unknown. Though he is usually pictured as a man in his thirties, he may have been older. While Luke says he was about thirty when he began his ministry, John says he was "not yet fifty," implying he was at least in his forties, maybe his late forties. Since no one knows for sure when he was born, it's impossible to say how old he was when he preached.

We do know when he died, maybe. By one account it was on Friday, March 30, in the year 36. But some people say he died in 29, some in 30, and others still in 33.

Whether he was resurrected is a matter of faith. But interestingly, we don't even know where the resurrection allegedly took place. Matthew and Mark imply it was in Galilee; Luke implies it was in Jerusalem.[6]

CHRISTIANITY

What of the red-letter dates of Christianity's development? These, too, have often been mythologized.

Take Constantine's conversion. This "turning point in the history of Christianity," wasn't one. We only think it was because historians needed a convenient date to mark the Christianization of the Roman Empire. In fact, by A.D. 312, the year Constantine had his famous vision on the Milvian Bridge, Christianity was well established. Besides, if he had a vision he didn't seem to take it all that seriously. He didn't agree to be baptized until just before his death. And he never relinquished his title as the official leader of paganism.*

Aside from his conversion, you never hear much else about Constantine, which is probably just as well. If the facts got out—that he murdered his rivals for the throne, persecuted pagans, and murdered his second wife**—he'd be hard to hold up as a hero of Christianity.

---

* There are still some historians who insist his conversion was sincere. Maybe he was just lucky, then, that it also happened to be politically beneficial.

** To be fair, I should point out that he killed his wife because she killed one of his sons.

It may be unfair to add, as well, that his sons were ruthless and brutal, but I can't help myself. After his death they killed off several cousins whom they considered potential rivals. Then they began killing off each other, taking thousands down with them until just one son remained.[7]

Another major event in Christian history, the Reformation, is also mangled.

Martin Luther is accurately remembered as the Catholic Church's leading critic, but his attack on the church is misconstrued. He never derided the whole church system. His chief objection was to the traffic in indulgences, the practice of selling pardons for sin. It was H. Zwingli in Switzerland who declared the whole system needed to be brought down.

Nor did Henry VIII break completely with Catholicism. He may have denied the supremacy of the pope, closed the monasteries, shut down the nunneries, and divorced several wives, but he never gave up his Catholic faith. During his reign he denounced Martin Luther, he approved of the persecution of Protestant reformers,* and in 1539 he insisted that Parliament pass the Act of Six Articles, which made it a crime punishable by death to deny transubstantiation, the Catholic belief that the bread and wine served at Holy Communion indicate the presence of Christ. He also insisted that priests remain celibate.[8]

Rich fathers with daughters were especially displeased with Henry, though it takes a little explaining to understand why. It seems that a lot of wealthy

---

* Several were burned at the stake.

fathers had used the nunneries as a dumping ground for their unmarried daughters. Now that the nunneries were closed, the daughters had to be married off. The problem wasn't that husbands now had to be found for the daughters. The problem was that when a rich man's daughter was married off a sizeable portion of the family fortune went with her. Too many such marriages and a family could find itself broke.[9]

Priests, by the way, weren't always celibate. Peter, the first pope, was married, as were numberless other Catholic clergymen. Not until the twelfth century did the church decide that priests shouldn't marry and shouldn't have sex.

Priests weren't required to abstain from sex because sex was considered bad or distracting. The ban was imposed to improve the average priest's status, which for some time had been in decline.[10]

Speaking of priests, the belief that Catholic priests the world over always read and spoke Latin is untrue. In England many priests in the medieval period were so unfamiliar with Latin they couldn't even construe the "opening words of the first prayer in the Canon of the Mass." In the diocese of Salisbury, for instance, five of the seventeen priests were illiterate in Latin.[11]

PART 11

WORLD WARS I AND II

WORLD WAR I

NAZISM

WORLD WAR II

HITLER

MUSSOLINI

CHURCHILL

HIROHITO

Hitler

WORLD WAR I

The easiest myth to dispense with about the First World War is that it was the first. According to historian Thomas Bailey, there have been at least nine world wars. By his count, the First World War was actually the Eighth World War.*

It's also thought to have been the first war in which the airplane was used. This isn't true either. The first plane flown in a war was in 1912, during the Mexican Revolution. It was piloted by a Frenchman, who flew bombing missions on behalf of the rebels. It's nice to know the Europeans got in some practical experience dropping bombs on others before they started dropping them on themselves.

---

* The others? (The dates refer to the years the contests became world wars.)

1) War of League of Augsburg (1688–1697)
2) War of Spanish Succession (1701–1713)
3) War of Austrian Succession (1740–1748)
4) Seven Years' War (1756–1763)
5) War of the American Revolution (1778–1783)
6) Wars of the French Revolution (1793–1802)
7) Napoleonic Wars (1803–1815)

Incidentally, though you hear a lot about aerial warfare in World War I, the size of the aerial fleets is usually underestimated. Factories turned out thousands upon thousands of planes during the war. England produced 58,000; France more than 68,000.

Of all the pilots in World War I, the greatest and best known was the Red Baron, Manfred von Richthofen, who was said to have shot down some eighty Allied planes. He himself was said to have been shot down in an aerial dogfight. It now appears, however, that he was killed in the air by a machine gunner firing from the ground.

Another first for which World War I is famous is that it was the first war in which chemical warfare was employed. So people believed at the time and so they believe now, but it's not true. Chemical warfare has a long history. Alexander the Great's army threw lime at enemy soldiers to distract them with burning and itching. The Barbary pirates burned opiates to create a lethal gas. The Spanish in the seventeenth century used smoke bombs filled with blood.

That war was inevitable because of entangling alliances is widely held but uncertain. Most of the nations which got into the fight would have fought anyway, alliances or no alliances. As A. J. P. Taylor has noted, they had no choice. Russia got into the fight because Austria invaded the Balkans, and if the Balkans fell, so could Constantinople and with it control of the passage to the Black Sea, Russia's economic lifeline. Germany had to fight once Russia got into the war. France got in the war because Germany invaded. And England couldn't stay out unless "she had been prepared to let Germany defeat France and Russia."

Who can be said to have started the war? Austria. Austria claimed it invaded Serbia in retaliation for the assassinations of Austrian Archduke Ferdinand and his wife. But this was madness. Prior to the invasion Serbia had met virtually all of Austria's demands. War came because Austria wanted war.

One of the memorable episodes of the war, which has helped confirm the lingering suspicion that the Germans started it, was the German chancellor's description of England's treaty with Belgium as "a scrap of paper." The quotation seems to indicate the Germans held civilization itself in contempt. But did he say it? The remark was supposed to have been made in the course of a heated argument between the chancellor and the British ambassador to Germany. But it's now believed the ambassador "gingered" up the remark in his report to the home office. According to an official British investigation after the war, what the chancellor actually said was that the treaty was "a piece of paper." No one knows for sure why the ambassador got the remark wrong, but it's possible he had in mind a play which had recently been staged at his own home and in which he had had a part. The name of the play: *A Scrap of Paper.*

That Germany and England ended up enemies in World War I is widely considered inevitable but wasn't. At the turn of the century, it seemed more likely that England and Germany would be friends than England and France, which shared a long history of enmity. Albert, Queen Victoria's late husband, after all, was German. The kaiser's mother was English.

Germany's difficulties after the war were long blamed by many, including Hitler, on war reparations required under the Versailles Treaty which ended the war. But the reparations could not possibly have seriously damaged the German economy. It turns out the payments Germany made to the Allies were far offset by the loans the Allies made to the Germans. Between 1919 and 1931 Germany paid 16.8 billion gold marks in reparations to the Allies. During the same time, private citizens in the Allied countries loaned Germany 44.7 billion gold marks. As historian Stephen Schuker has observed, "The net capital flow thus ran strongly toward Germany." By his calculations, "the inflow amounted to approximately 2 percent of national income during the entire Weimar Republic. Some vengeance! Some ruin!"

Whether the Versailles Treaty was harsh or not, the Germans were hardly in a moral position to complain. When they had made peace the previous year with the new Russian government, they made far greater demands on the Russians than the Allies subsequently made on them. By the Treaty of Brest-Litovsk the Russians were made to give up Lithuania, Latvia, Estonia, Russian Poland, and several strategic islands in the Gulf of Riga. In addition, Russia had to agree to the independence of Finland, Georgia, and the Ukraine. Cut off from the Black Sea, the Russians also were made to pay reparations of six billion marks.

Too much, in any case, has been made of Germany's weakness after the war. Germany in defeat was stronger than France in victory. France, to be sure, had won reparations and the return of Alsace and Lorraine, but the French had suffered the highest casualty rates of any of the warring powers. Figures show that seventy-three percent of the mobilized French forces were killed in the war. (That is: 1.4 million killed.)

And Germany? Germany lost even more people than France (1.8 million), but the German population after Versailles was twice as large as France's. And most of the war had been fought in northern France. German territory was largely unaffected. Finally, Germany emerged from the war in a stronger strategic position than it had been in before. To the west, the Germans now faced a weakened France; to the east, Russia (which was in chaos) and the Slavic states (which were now divided).[1]

NAZISM

People know a lot about the Nazi era, but much of their knowledge is based on information gleaned in World War II, and much of the information available then was misleading or inaccurate. Everybody's heard, for instance, that the Germans turned to Hitler in part because of inflation, which was so bad people needed a wheelbarrow of cash just to buy a loaf of bread. But Hitler did not stop inflation. He didn't have to. The Weimar Republic had stopped it.[2]

The story about the Reichstag fire is that the Nazis started it. But there's no evidence they did. None. According to historian A. J. P. Taylor, it was a Dutch socialist named Marinus van der Lubbe who started the conflagration. Van der Lubbe, who did it to protest Nazism, confessed to the crime, was prosecuted for the crime, and was found guilty of it at a fair trial. Following the trial, his head was chopped off with an ax.

Why is it commonly thought that the Nazis did it? Well, for one thing, it certainly seemed like them. For another, they stood to benefit from the fire, as it gave them an excuse to exploit people's fears of social chaos, which, being Nazis and all, they did rather well. At the general election, held just a week after the fire broke out, the Nazis won a working majority in parliament.

Most history textbooks still insist the Nazis themselves were behind the blaze, because the authors cannot believe that Hitler merely made the most of an incident of someone else's doing. But it would seem that he did.[3]

That Nazism represented a clear ideological break with the German past is widely believed but unsubstantiated. Take Nazi history books. According to E. H. Dance, "Most of the Nazi teaching about history was not Nazi at all. It is German—and it can be found in German history books published long before Nazism was born." Even the history books used under the liberal Weimar Republic were compatible with Nazism, teaching that Germans are racially superior, that Germany had to fear encirclement by its neighbors, that Germany was destined to dominate Europe, and that Germans should revere the principle of leadership, *Fuhrertum*.[4]

That the Nazis were militaristic because Germans by nature are militaristic is so well established that nothing is apt to dislodge the idea. But German militarism is exaggerated. Through most of history Germans were not regarded as any more aggressive than their neighbors.

Behind the fear of German aggressiveness lies the fixed stereotype of the fierce Prussian soldier, but the Prussians are gotten wrong. While Prussians long took pride in their army, which helped unify the state, the army did not become fearfully aggressive until Bismarck, and nobody much worried about it until then. Consider world reaction in the eighteenth century to Frederick the Great's rape of Silesia. Europe condemned it, but as Michael Howard shrewdly observes, "this stroke of *Machtpolitik* was not considered by contemporaries to display any peculiarly *Prussian* characteristics." If anything, says Howard, it was decidedly un-Prussian-like, Frederick's forebears having achieved a deserved reputation for docility.

Bismarck's militarism is undeniable, but Bismarck did not rule in the name of the old Prussian leadership. Indeed, the old Prussian leaders, the Hohenzollerns, were regarded as stumbling blocks to Bismarck's plan for a grand German state.

Nazis eventually won the support of many of Prussia's leaders. But the early Nazis weren't Prussian. Their roots were in southern Germany, not Prussia.

It could be argued that the problem with the Nazis is that they weren't Prussian enough. As Howard observed, "the old Prussian virtues *were* virtues and remain so. Industry; piety; frugality; self-discipline; physical courage; a capacity for self-subordination to a common cause: without these no society can survive, whatever its political complexion." And Prussians, historically, were tolerant of Jews.[5]

The "ruthlessly efficient" Third Reich of which we hear so much was indeed ruthless, but it wasn't efficient. The only reason we think it was is because the Nazis said it was. But the Nazis were lying. Historians now report that the Reich was continuously divided by factionalism and often paralyzed by internal conflict. "Behind the goose-stepping columns and the facade of order," says V. R. Berghahn, "there reigned administrative chaos and anarchy." Even Hitler's legendary grip on the government has been questioned, one historian concluding he was a "weak dictator."

Evidence? During the war the Nazis were never able to put the economy on a wartime footing. The Germans never had enough arms, never enough planes, never enough supplies. The reason they launched blitzkreig attacks was because they couldn't sustain attacks that lasted longer. Britain, though it had a smaller national income, regularly outproduced Germany. In 1941 it turned out twice as many aircraft, a thousand more tanks, and five thousand more big guns. The Nazi government exercised so little control over the economy that officials could not even stop families from hiring domestic help, though it was a huge and unnecessary expense. In 1944 Germans employed 1.3 million servants.[6]

Of all the dark deeds of the Nazis none was more reprehensible, of course, than the systematic campaign to exterminate the Jews, but the idea that Hitler always intended to exterminate the Jews is now open to dispute. While no one doubts the lifelong intensity of his anti-Semitism, some now suggest that in the early years of the Third Reich he may have been content to deport them. Mass deportations were common in the thirties. By 1938 the Nazis had deported nearly 200,000 Jews from Germany.

Why didn't Hitler, then, simply continue deporting Jews? Because he couldn't find a place willing to accept them. Britain even backed out of a plan worked out by the SS to deport Jews to Palestine. In 1940 Hitler personally endorsed a proposal to deport all remaining Jews in Germany to the French island of Madagascar, off the coast of Africa, but the plan proved impractical. Because the British controlled the sea lanes, it couldn't be carried out.[7]

WORLD WAR II

Everybody knows what happened in World War II, having seen the movies about it on late-night television. BUT DAMNED IF MYTHS DON'T REMAIN!

Take Munich. A disaster, right? But why? Since when is a peace conference that actually produces peace a disaster? Yet from what's said you'd think the peace conference of 1938 was the most horrible event in the history of the world.

The assumption people make is that it would have been better in 1938 for Great Britain to confront Germany than to appease it. But in 1938 Great Britain could not risk confronting Germany. Great Britain was weak militarily and needed time to rearm. When war broke out a year later it was in a far better position to fight. The belief that the British lacked the guts to confront Hitler in 1938 is groundless myth.

Besides, Great Britain wasn't the world power it's believed to have been. Though the British flag flew over a quarter of the globe, as the British liked to boast, the empire was in tatters and in economic decline. By 1900 it had been eclipsed by the United States as the world's leading industrial power. By the 1920s it was no longer even the world's leading financial power. (Forty percent of the budget went to paying off the interest on the national debt.) As historian Clive Ponting observes, by the 1930s Britain was "attempting to control about a quarter of the globe with only ten per cent of its manufacturing strength."

Besides, as the British military warned Neville Chamberlain, the empire had too many enemies. In 1937 the British chiefs of staff concluded: "We cannot safeguard our territory, trade and vital interests against Germany, Italy and Japan simultaneously." Given this situation, what the British sensibly tried to do was reduce the number of potential enemies they faced by appeasing one of them.

Appeasing Germany in 1938 was not the key mistake, therefore. The mistake was allowing Germany to become so powerful earlier that Britain and France had no choice by 1938 but to adopt a policy of appeasement.*

Probably the opponents of appeasement are right to think that "you can't appease a dictator." But sometimes you don't have much choice.

Nobody today, of course, approves of the sellout of Czechoslovakia. But would it have been morally superior to fight and lose?

---

* Maybe people like Churchill were right to attack appeasement as the policy of defeatists. But even Churchill, when he first came to power, tried appeasement, too. See p. 253.

Hitler, incidentally, disliked the Munich agreement, which indicates maybe it wasn't all bad. Though it gave him what he wanted at virtually no cost, the agreement was (says historian D. C. Watt) "imposed on him and he deeply resented it."

To be sure, Munich did not bring "peace in our time," as Neville Chamberlain proclaimed. But Chamberlain wasn't the only one who approved of the deal. So did most of the world's leaders, including Franklin Roosevelt. It's forgotten that Roosevelt not only supported the Munich agreement, he tried to take credit for it.* Roosevelt, indeed, had supported appeasement from the time Hitler took power. In 1933 he told his roving envoy Norman Davis that "political appeasement" was needed to provide for lasting peace. Later that year he told the German ambassador to the United States that Hitler was the right man to lead Germany. In 1935 he asked an old business friend, Samuel Fuller, to find out what the Germans wanted in exchange for peace. In 1937 he sent Sumner Welles to Europe to see if Britain would agree to the return of Germany's African colonies. In 1938 he fired William Dodd, the United States ambassador to Germany, after Dodd made some speeches attacking the Nazi Party. And when the Germans in 1938 demanded the western part of Czechoslovakia (the Sudetenland) as the price for peace, Roosevelt gave them his support; thus, the Munich Agreement.**[8]

* Fortunately for him, he was denied the honor.

** It's true that Roosevelt gave speeches in the thirties decrying the tyranny of the European dictators. But he also gave speeches in which he said that the way Europeans governed themselves was their concern, not ours.

Roosevelt also favored the appeasement of Italy. This isn't well known, but it's true. And he continued appeasing Italy right up until Italy went to war against France, when he finally stopped. This would seem to have been a little late. It was after Germany had already taken over Czechoslovakia, after Germany had invaded Poland, after Germany had conquered Norway and Denmark, after Germany had invaded Holland, Luxembourg, and Belgium, and after Dunkirk. Perhaps it was worth trying to keep Italy out of the war, but FDR maybe kept trying a little too long.

The popular explanation of the quick defeat of France is that the French were unprepared. This is partly true. But the French were not nearly as unprepared as people say. They had more tanks than the Germans and their tanks were better. They also had plenty of planes, when you count the squadrons contributed by Britain. As William Shirer reported in his history of the French defeat, "they had enough aircraft to give the Germans a good deal of trouble."[9]

Much has been made of the German tank divisions which overran France in six weeks. Much too much. The French collapsed because they lacked the will to fight, and because they had expected the Germans to strike through Holland and Belgium as they had in World War I and not through the Ardennes. Hitler was going to oblige them until his plans accidentally fell into Allied hands, inspiring him to change his strategy just months before the invasion.*

Besides, the much-feared Panzer divisions are overrated. Clive Ponting reports that "only five per cent" of the German army "was in armoured Panzer divisions and ninety per cent of the tanks in those divisions were obsolete training models dating from the early 1930s or taken over from the Czech army in 1939." Relatively few German tanks (the Mark III and Mark IV) "matched Allied models and Germany produced just forty-five Mark IVs in the whole of 1939." By the end of the war the Germans were producing some 1,500 Panzers a month, but in 1940 they possessed just 2,500 tanks in all, a thousand fewer than the Allies (not including the United States).

If too much is made of the Panzer divisions, too much is also made of the German army as a whole at this time. Everybody who's examined the army in detail has reached the conclusion it was inadequately trained and ill-equipped. Little time had been allowed for the expansion of the army, which had increased to three million men (from 100,000) over just a brief five-year period. And insufficient

---

* You can drive yourself crazy playing the "what if" game, but it's almost impossible not to. What if Hitler had not changed his strategy, as he almost didn't? Would the Maginot fortifications have stopped him? What if his plans had not accidentally fallen into Allied hands, which was the result of a plane crash in Belgium? Would he have stuck to his original strategy, or was he uncomfortable all along repeating the approach taken in World War I? You can't change history, but it's worthwhile remembering it didn't have to turn out the way it did.

efforts had been made to provide the army with supplies. The movies have left the impression, for example, that the German army moved through Europe at lightning speeds in sleek tanks and modern trucks. But half the army still moved about on horses. The army used 2.7 million horses during the war.

The Poles also used horses, but everybody knows that. That's usually given as one of the reasons they lost so quickly. Actually, the Germans used horses in the Polish campaign, too: some 200,000 horses.

As far as horses go, the United States was far behind. The U.S. Army in 1941 had only about 50,000 horses. But this deficiency didn't seem to amount to much.*

The miracle of Dunkirk is not that the British were able to get their soldiers out of France safely. That was expected, given that French and Belgian troops were willing to hold off the Germans until the British were able to complete the evacuation. The miracle is that the retreat, which was obviously a defeat for the British army, came to be regarded almost as a heroic victory. In fact, the retreat of the British Expeditionary Force from France in the spring of 1940 was (as Churchill confessed, in private) "the greatest British military defeat for many centuries."

As might be expected, given the conditions, it was not even carried out well. At Calais, when Ger-

---

* Never heard that the Nazis used horses in World War II? Never knew the U.S. did as well? Neither did most contemporaries. The widespread use of horses in the war was revealed as a news item in 1941 in *You're Wrong About That* (May/June 1941), pp. 5–8.

man shells began landing, British stevedores had to be forced to work. And at Dunkirk some officers abandoned their positions to catch the earliest boats out. Things became so chaotic at Dunkirk that British sailors had to resort to armed threats to keep the troops from storming the ships. Upon returning to England the troops were so demoralized, according to an official in the War Office, that "they threw their rifles and equipment out of railroad carriage windows." He concluded: "The Dunkirk episode was far worse than was ever realised in Fleet Street."

In the weeks leading up to the retreat the French had demanded that the British put up a fight, but the British military had refused. Churchill had to issue a direct order to Lord Gort, the commander of the British army on the coast, to get him to fight. Churchill, in disgust, telegraphed Gort, "Of course if one side fights and the other does not, the war is apt to become somewhat unequal."

The French do not regard Dunkirk the way the British do, and who can blame them? When the British began evacuating their troops they refused to allow French forces to go with them. The French were not even officially notified that the British intended to retreat from Dunkirk until two days after the withdrawal began. Not until then were French troops in large numbers allowed to join their British allies.

Worth noting in passing is that the great majority of troops were evacuated in Royal Navy ships. Only eight percent were evacuated in the legendary armada of little boats sailed by volunteers.

I come now to the Battle of Britain and the Blitz.*

No doubt these twin episodes of destruction deserve to be remembered as they are, as Britain's darkest days during World War II. As the movies have accurately shown, the English suffered greatly during this period. More than 40,000 were killed. More than 86,000 were injured. And over two million homes were laid waste.

But if the British people suffered they did not suffer in silence, nobly or otherwise, and they did not take the attacks in stride. They were bitter and they were demoralized. The only reason we think they weren't is because of a vigorous British propaganda campaign. The government knew better. It was secretly opening the people's mail. The mail operation, employing ten thousand snoopers, revealed that the nation was in a panic. Rumors circulated that the royal family itself had fled to Canada for safety. The Ministry of Information, after reviewing the reports filed by its snoopers, concluded: "Public morale is at a low ebb."

From the newsreels and from Edward R. Murrow's electrifying London broadcasts, it's usually thought that during this period the English mainly spent their days and nights huddled in prayer in darkened basements. They did, during the raids. But the rich, when the bombs weren't dropping, stuck to a scandalously lavish social schedule. This is the story Ed Murrow left out. And what a story it

---

* The Battle of Britain took place in 1940, from July through September. The Blitz began in September 1940 and lasted until May 1941.

would have made. Take, for instance, the social calendar of John Colville, Churchill's secretary. In August, the week of the worst bombing in London, he attended two society lunches, ate two dinners at a top restaurant, went to the theater and a nightclub, and even found time to visit a country estate and play tennis. Churchill dined on caviar and oysters, though he was inconvenienced by having to sleep in a bunker. The rich drank champagne, danced, and held fancy dinners.

Occasionally, of course, the war interfered with the wealthy's schedule of regular meals. When this happened, they complained. Diplomat Harold Nicolson, required to take meals at the restaurant at the Ministry of Information, griped: "It is absolutely foul. It is run on the cafeteria system and we have got to queue up with trays with the messenger boys."

The great majority of the English suffered just as we think they did. But war hardly brought out the best in people. Under the strain, many turned to crime. Beginning with the Battle of Britain the crime rate soared, eventually climbing by sixty percent. So many became thieves that Scotland Yard, according to Clive Ponting, "had to set up a special anti-looting squad." Almost half of those caught were civil defense workers.

The survival of Britain during the Nazi air attacks is usually attributed to the heroic efforts of the pilots of the Royal Air Force. But the country's survival was actually due to radar, which the British adopted on the hunch of the government's science advisor, a bookish man named Henry Tizard, of whom little is ever said and who, shortly after his triumph, was forced out after losing a bureaucratic fight with one of Churchill's closest advisors.

To be sure, the pilots of the RAF deserved the accolades Churchill bestowed on them. (It was them he had in mind in the speech in which he remarked, "Never in the field of human conflict was so much owed by so many to so few.") But their legendary feats are exaggerated. Only fifteen percent of the RAF's pilots ever shot down a single plane. And just seventeen pilots shot down more than ten. And it wasn't the RAF's English pilots who made the most kills. It was an RAF Polish squadron. Of all the hundreds and hundreds of RAF pilots, the two most successful were a Czech and a Pole.

Hitler, incidentally, never planned on bombing London or any other major population center. As late as August 24, 1940, two weeks before the start of the Blitz, he expressly forbade the bombing of London. He relented only after the British had bombed Berlin. The British attack came in response to the Nazi bombing of a London suburb. The Nazi bombing of the suburb, however, had been a mistake. The Nazis had been trying to take out oil tanks in nearby Kent.

The United States, it is worth noting, did not believe the British could withstand the attacks of the Luftwaffe. When Churchill asked Roosevelt for destroyers, Roosevelt at first refused, on the grounds that England probably would fall as France had.

Eventually, Roosevelt agreed to the celebrated Destroyer Deal, which gave Britain fifty aging destroyers in return for 99-year leases to bases in Newfoundland, the Caribbean, and Bermuda. But contrary to popular impression, the deal was not made so much to help Britain as to help the United States. Military advisors had told the president that the United States desperately needed those bases. The British felt the deal was decidedly unfair, but they desperately needed those ships.

In the end, Roosevelt agreed to the deal only if Churchill would promise to sail the ships back to North America (along with the rest of the British fleet) in the event of a successful German invasion. Churchill agreed, but hadn't wanted to. If the Germans invaded, he had expected to use the fleet as a bargaining chip in negotiations.

Another deal made with the British, Lend-Lease, is equally misunderstood. Roosevelt told the public the goods loaned to the Allies would be returned or paid for. Most weren't, however. Under Lend-Lease fifty billion dollars in military goods were loaned out; only ten billion dollars in goods were ever returned. Great Britain received 31 billion dollars in goods and paid back only 650 million. The isolationists had predicted this would happen. It was the only thing they were right about. (Isolationist Robert Taft had commented that loaning arms was like loaning chewing gum: "You don't want it back.")[10]

About Pearl Harbor, it's hard to know what to believe anymore. Anyone who's confused has a right to be. Formerly, for instance, it was held by some that Roosevelt had deliberately allowed the bombing of the island in order to bring the United States into the war. Now it's claimed that if Roosevelt didn't have advance knowledge of the attack, Churchill did.

James Rusbridger and Eric Nave* argue that British intelligence, having cracked the Japanese naval code, knew as early as November 26, 1941, that the Japanese fleet had left its home port, and knew as of December 2 of the message, "Climb Niitakayama 1208," a reference, presumably, to an attack scheduled December 8 (Tokyo time). British intelligence officials are said to have guessed that an attack somewhere was imminent and that it was going to occur at one of several specific locations (the Philippines, Dutch East Indies, Singapore, or Pearl Harbor). But whether Churchill knew what they knew and whether he deliberately withheld the information from Roosevelt is undocumented. (The book was dismissed by Herbert Mitgang in the *New York Times*. But it was chosen as a selection of the History Book Club and has been praised by professional historians.)

Whether the attack should have come as a surprise or not, it did. And the Japanese wanted it that way. But they did not, it should be pointed out, plan on the attack coming while the United States and Japan were still officially at peace. A half hour or so before the attack the Japanese were supposed to notify the U.S. that they had declared war. They didn't, however, because of bureaucratic delays at the Japanese Embassy in Washington. (But would it have made any difference? The Japanese seemed to think so. But would Americans?)

All know that Pearl Harbor was a disaster:

---

* In *Betrayal at Pearl Harbor* (1991).

18 ships sunk, 347 aircraft destroyed, 4,000 casualties. But it was not a strategic catastrophe. Stanley Weintraub has noted that the eight battleships that were lost were obsolete and so were half the planes. What would have been catastrophic would have been the sinking of any of the country's aircraft carriers, but none was sunk. They weren't there.[11]

Another myth that came out of the war concerns German P.O.W.'s. Everybody knows German soldiers rushed to surrender to the United States Army at the end of the war because they were afraid to surrender to the Soviet Union. What few realized until recently, however, was that they did not fare well even in American hands. It's now been revealed that at least 56,000 German P.O.W.'s died while in American custody, almost all, probably, from malnutrition. It's the best-kept secret of World War II.

The American army did not purposely starve the Germans. There simply wasn't enough food to go around to meet the needs of both German civilians and German P.O.W.'s. When shortages developed, the army made sure the civilians got fed before the P.O.W.'s did. Conditions in the American P.O.W. camps were so terrible that General Eisenhower reclassified the P.O.W.'s as D.E.F.'s—Disarmed Enemy Forces—to exempt them from the minimum standards of care required for P.O.W.'s under the Geneva Convention.

It's been charged that up to a million German P.O.W.'s may have died in American custody. A panel of eminent historians found that this was not true. But they did conclude that "there was widespread mistreatment of German prisoners in the spring and summer of 1945. Men were beaten, denied water, forced to live in open camps without shelter, given inadequate food rations and inadequate medical care."[12]

HITLER

And who was responsible for World War II, a war which:

- Left fifty million dead
- Reshaped the destiny of the Jews
- Remade the map of Europe
- Led to the development of the first atomic bomb
- Destroyed the British Empire
- AND gave birth to the Cold War???

An asexual paperhanger and house painter, Adolf Hitler.

Only he wasn't a paperhanger: that was just a silly story put out by the Allies to discredit him.

And he wasn't ever a house painter. He had been a regular old painter, the kind who paints pictures.

And he wasn't asexual. The Hitler who stoically devoted himself entirely to the Nazi cause is the Hitler of myth. Although he didn't drink and barely ate, he *was* interested in sex. He took naked pictures of women, invited striptease dancers to perform naked at small gatherings, and had numerous affairs.

Glenn Infield, author of *Hitler's Secret Life,* estimates that Hitler as an adult had affairs with at least eleven women, including an Italian countess, a German actress, a niece, a sixteen-year-old, Richard Wagner's daughter-in-law, and Hermann Goring's wife, and insists the Fuhrer had sex with many of them. (It's suggested Hitler may have even had a child by Goring's wife.)

Infield's evidence, because it largely consists of statements made by the women themselves, is considered unreliable by many. But too many of the statements tell the same story—that Hitler was a sadomasochist—for them all to have been manufactured. The actress recounted how, after they undressed, he liked her to kick him. The niece was quoted as saying he liked her to sit on his face and urinate. Wagner's daughter-in-law confided (according to her daughter) that Hitler enjoyed being whipped.[13]

Why we decided it was better to think of Hitler as asexual, I don't know. But we did, in error.

That Hitler is regarded today as a fanatic is due to the fact that he was one. But the idea that the Germans elected him because he was a fanatic is wrong. Often Hitler concealed his fanaticism, leading most Germans to think of him as a moderate. In 1936, for instance, when the Nazi Party took on the Catholic Church, demanding the removal of crucifixes from church classrooms, Hitler stood aloof from the controversy, leaving intact his reputation as one of the "good God-fearing Nazis." Two years later, following Crystal Night, the Nazi terrorist smashing of Jewish shops that ended with the imprisonment of 20,000 Jews in concentration camps, Hitler again publicly distanced himself from the radicals. Everyone in Germany knew he hated Jews, but he tried to pretend he wasn't one of the crude Jew-haters. Not Adolf Hitler.

Not yet, anyway.[14]

MUSSOLINI

What of Hitler's pal dictator, Mussolini? He wasn't as tough as they say he was. As a young man he feared walking home alone at night. His friends uniformly described him as timid. He is said to have lived in fear of his wife. And he took stress badly, so badly he dieted on milk of magnesia to relieve the symptoms of an ulcer (though an autopsy proved he'd never had an ulcer).

He was also the superstitious type. His biographer says he was "terrified of the evil eye," and out of superstitious fears never personally signed a death warrant.[15]

The odd thing about Mussolini is that it took Americans a long time to get wise to him. In 1922, when he stormed Parliament and destroyed the Italian republic by force, the U.S. ambassador quaintly commented: "We are having a fine young revolution here. No danger. Plenty of enthusiasm and color. We all enjoy it."

When in 1924 news reports directly linked Mussolini to the kidnapping and murder of a leading Socialist, Giacomo Matteotti, the United States blithely ignored the story; officials said it wasn't important. After all, Mussolini was keeping Italy from going communist and his people seemed to like him, so who were we to object?

In 1931, when Secretary of State Henry Stimson visited Italy, he praised Mussolini for establishing order and suggested that fascism might be just what Italians needed. "Americans could understand from their frontier experience," said Stimson, "that in a time of lawlessness there might be a need for vigilantism."

In 1933 Roosevelt told friends he was "deeply impressed" with Mussolini, whom he described as an "admirable Italian gentleman." And Roosevelt continued to express confidence in Mussolini even after an American general reported that the dictator had run over a young girl and didn't care. ("What is one life in the affairs of a state?" Mussolini is supposed to have remarked.) Roosevelt eventually came around to the idea that Mussolini was a madman like Hitler, but it wasn't until Mussolini invaded Albania.[16]

CHURCHILL

Winston Churchill was the popular leader of England who condemned appeasement and gave a lot of stirring speeches during World War II. Only he wasn't always popular, he didn't always condemn appeasement, and his greatest speeches he himself didn't broadcast.

Popular? The only reason he got the job of prime minister was that the preferred candidate, Lord Halifax, had refused to take it.*

Churchill, his first time in the cabinet, during World War I, hadn't done too well. It was Churchill who'd been largely responsible for the Gallipoli disaster in which 55,000 Allied soldiers had been killed. The English found it hard to forget that.

Why then did they give Churchill a second chance? I think they were desperate.

Once in office, Churchill proved to be every bit

---

* Neville Chamberlain was of the opinion that Churchill wouldn't last and that he himself would be returned as prime minister. Chamberlain was wrong about lots of things.

as different from Chamberlain as people think. Whereas Chamberlain had publicly supported appeasement, Churchill was extremely careful only to do so in private.

Churchill in his own history of the war never mentioned that he'd approved of appeasement. I think this was because it wouldn't have looked so good.

But he did. At the very same time that he was publicly assuring the English that he wanted "victory at all costs," he was privately telling the cabinet he'd consider giving Germany back some of its African colonies and giving Italy Gibraltar, Malta, and the Suez Canal if that would get Britain "out of this jam."

Which reminds me: you know all those speeches he gave, the speeches with which he boosted the morale of the English people, the speeches in which he spoke so sonorously it was almost like he was making love to the microphone? Well, Churchill never actually delivered them over the radio. A stand-in did.* Churchill, having given them once in Parliament, did not want to waste time giving them again over the radio.

Which speeches? The one about "blood, toil, tears and sweat"; the one in which he exclaimed, "we shall fight on the beaches"; and the one about "their finest hour."

The "blood, toil, tears and sweat" speech, incidentally, was not terribly well received, in Parliament at least. Historians say the M.P.'s gave it a "cold reception."

---

* Norman Shelley, an English actor.

Nor was it Churchill who first made use of the expression "tears, sweat and blood." Churchill stole it from Byron. Byron stole it from John Donne.*

And speaking of stolen expressions: Churchill did not coin the line about the "iron curtain." Joseph Goebbels used it in 1945, Lady Snowden in 1920, and Queen Elisabeth of Belgium in 1914.[17]

---

* Well, to be honest, I don't know who Churchill actually stole it from: Byron or Donne. But he stole it.

HIROHITO

Hirohito started off well enough. He announced when he ascended the throne, in 1926, that he was taking Showa as his official name. In Japanese the name means "peace and enlightenment." After ward, this would strike almost everybody as grimly ironic. But at the time it was a perfectly good name.

As a young man he went off to England and visited with the Prince of Wales (later Edward VIII). Ever after he always insisted his stay in England was one of the happiest periods of his life. When he later declared war on England he said he was really sad. It always made him sad to turn on a friend. He would remain sad a whole long while, then little by little, he'd get better and forget all about it.

In the 1930s he was very busy. First, Japan made war on China. Then Japan took over French Indochina. Then Japan invaded Malaya. And finally, Japan went to war against the United States and Great Britain.

Hirohito claimed after World War II was over that if it had been up to him Japan never would have gone to war against anybody. But that's not how he seemed to feel at the time.

Take the war with China. While it's unclear whether he supported it in the first place, he made no effort to end it. In fact, in 1938 when his generals said they wanted to make peace with China so they could concentrate on a possible war with the Soviet Union, he insisted on it going forward. It would look bad, he said, to sue for peace; Chiang Kai-shek might get the idea he'd won.

In a way he did oppose the war. He told his generals that if they couldn't finish the war as quickly as they had promised (six months), then they shouldn't have started it in the first place. But I'm not sure this puts Hirohito in the peace camp.

Nobody's ever implicated him in the Rape of Nanking, but it's hard to believe he remained unaware of it for years, as he always maintained. His diplomats knew, his prime minister knew, his generals knew. How is it he didn't know? Of course, it's always possible he was deliberately shielded from information about the atrocities, but there's no evidence he was.

If he did not know about the Rape of Nanking, it was about the only action of the war of which he was not aware. We've all been taught to believe he remained detached from the day-to-day conduct of his wars, but he wasn't. He had a war room in the palace—known as the Grand Imperial Headquarters—where he monitored the movements of all of his troops.

He had mixed feelings about the invasion of French Indochina. He felt it was a little bit unfair to mount an invasion of the French colony just when the French had been overrun by Hitler and couldn't defend it. "But," he said, "I suppose it can't be helped."

His main concern with the invasion of Malaya was that it go off well. Success depended on Japan's getting its troops through Thailand without the European powers knowing it. Hirohito knew this was critical and kept asking his generals if Thailand could be counted on to keep things secret. He nagged them about this so much they finally wished he'd just go away.

He felt like most people in Japan probably did about World War II. He liked it when things were going well and he liked it less when things weren't. Through the years he liked it less and less until by the time of the American occupation it seemed he had never liked it at all.

After the war Hirohito said he'd opposed it "at every turn," but this wasn't true. The only time he expressed any reservation about a war with the United States was when he was told the navy wasn't sure it was ready for one. Hirohito worried a lot about the navy's preparedness. He told friends he only wanted to go to war if he knew Japan could win.

Pearl Harbor, on reflection, doesn't seem like such a good idea. But Hirohito liked it. Yes, he knew about the attack on Pearl Harbor before it happened. And he knew it was supposed to be a surprise. At no time did he express any reservation about launching a sneak attack. When word was radioed back that the attack had been a success, he celebrated.

Hirohito liked to pretend he would have prevented the war if he could have, but he said he couldn't. In September 1945, at their first meeting, General Douglas MacArthur asked Hirohito how it was he'd had the power to stop the war but hadn't had the power to prevent it. Hirohito answered that if he had tried to prevent the war he probably would have been assassinated or imprisoned in an asylum. I think this was supposed to explain his inaction. I'm not sure it does, though.

If we tend to think of Hirohito as the helpless puppet of the militarists, it's because MacArthur ultimately decided it would be helpful to the United States for people to think of him that way. And so we have. But Hirohito was his own man. In 1936 when army militarists tried to take over the government Hirohito personally led the crackdown against them, calling in the navy for help. Later, over the advice of his top assistants, he personally ordered the execution of the ringleaders of the rebellion.

I know he always looked shy and retiring. This was because he often was shy and retiring. But even shy and retiring types can act decisively every once in a while. No one wants to believe that the little marine biologist in glasses ever lied, but he did indeed lie about his involvement in World War II.

Nobody ever had any evidence he'd committed any war crimes. This, however, may have been because it's not easy to produce evidence that's been burned. And we know the Japanese engaged in the wholesale burning of documents just prior to the arrival of the Americans.

A case can be made that he must have known about the notorious activities of army unit 731, the unit that conducted germ warfare experiments on live humans. It was the only army unit established expressly by imperial decree and his youngest brother was one of the unit's officers.

There was one way the questions about his wartime record probably could have been answered: by examining his personal diary. But it was never subpoenaed. The war crimes prosecutors said they wanted it, but MacArthur said no and when he said no, he meant it.

Hirohito died in January 1989. It was just in time. A short while later Edward Behr published the book that exposed Hirohito's wartime past.[18]

PART 12

HOLLYWOOD DOES HISTORY

BASED ON A TRUE STORY

SOLDIERS AND WAR

NEWSREELS

Cleopatra

BASED ON A TRUE STORY

Say what you will about the movies, it is the movies that have done the most to inform people about world history. Then again, it is the movies that have also done the most to misinform people about world history. It hurts to say so, but most history is not half as interesting, romantic, or simple as Hollywood makes it out to be.* Nor, for that matter, were the heroes and heroines of history usually as attractive as the stars who've played them.

Hairstyles, to begin with the basics, seem to have been an especially difficult thing for Hollywood to get right. Consider Cleopatra's coiffure. Anybody who's seen the movies about her probably thinks she wore bangs, for in the two biggest movies made about her the actresses playing her wore

* The purists always insist that the truth is at least as interesting as Hollywood's fictions. And maybe it is. But moviegoers seem to have a different idea. For some reason they seem to show a marked preference for pictures that feature blood, sex, romance, and spectacle in huge, extravagant dollops.

bangs: Claudette Colbert in Cecil B. DeMille's 1934 classic, and Elizabeth Taylor in the 1963 remake. But Cleopatra never wore them. She wore a wig with tight curls over a shaved head. The reason Claudette Colbert wore bangs is because she apparently had a personal fondness for them. Elizabeth Taylor wore them because in the early 1960s bangs were in.*[1]

Hairdos of the pharaohs are almost always done right. But the movies never show them with hair on their face, though almost all of them wore so-called beard-wigs, which extended from the chin like a goatee in a long braid. Even reigning queens wore them.

What kind of hairdos the ancient slaves wore, it's difficult to say. But I understand they did not go around with a flat-top as Kirk Douglas did in the 1960 Stanley Kubrick classic about Spartacus.

Historians know exactly how Elizabeth, the Virgin Queen, looked: she was completely bald, lacking even eyebrows and eyelashes, lost in an early illness. But only once has an actress been allowed to play the part bald. That was in the 1956 film starring Bette Davis.

Marie Antoinette, the hated French queen, is always featured wearing a white wig, as are all of the ladies of the French court in the eighteenth century. But they didn't wear white wigs, they wore gray wigs. Hollywood made the actresses wear white, however, because on the silver screen white wigs look more elegant.

---

* For more myths about Cleopatra see pp. 19–23.

Moviemakers do go to great lengths to make sure that the clothes worn by actors in history epics are authentic. Like the time a director made sure that even the actresses' petticoats (unseen by the audience) were exact duplicates of the originals. Some cynics snarled that it was all just a publicity stunt. I'm sure the critics were in error.

But most of the time the clothes in costume movies have reflected the fashions of the era in which the movies were made. In *Marie Antoinette,* for instance, Norma Shearer was put in gowns that revealed her bare shoulders, because the year the movie came out (1938) bare shoulders were hip. In the 1930s, when designers began cutting fabrics for women on the bias, to fit the curves of the body, Maid Marion and other women from history suddenly started showing up on screen with clothes cut the same way as well. In the 1950s, when the "lift and separate" bra became available, they began sporting modern bustlines. It's a little shocking to think that through most of history famous women had to get by with loose-fitting dresses and flat bosoms, but thanks to Hollywood, this is one shock the American public has not had to face.

Until Shakespeare's day dramatists didn't even attempt to dress their actors like the historical figures they were supposed to be playing. Why bother, they felt?

Say you were going to put on a production featuring Cleopatra. Put her in some Egyptian outfit? Why?

So you know how they'd dress her? As she was

a queen and all they made her look like Queen Elizabeth.[2]

By the time of Shakespeare, dramatists began providing more accurate costumes. But as historian Quentin Bell reports, figures like Cleopatra and Macbeth were still dressed "as though for a box in the auditorium."

Moviemakers typically have taken into account not just prevailing fashions, but also the audience's interest in . . . SEX! Take Josef von Sternberg's never-released movie about Claudius. If he had wanted to be accurate, he would have dressed his vestal virgins demurely and he would have featured just six of them, as that's all there were. But he put them in "costumes resembling bikinis under gauzy drapery" and he put in SIXTY of them. As he told the costume designer, "I want sixty, and I want them naked."[3]

Or take the ancient Greek muscleman movies of the 1950s. There's only one reason the Greeks in those movies had big bulging muscles. It's because Hollywood discovered that ancient Greeks with big muscles sell tickets. Real Greeks back in the days of Hercules didn't have big muscles. They had bodies that were well-proportioned. Big muscles didn't come until Alexander's day. So Hollywood's given us all the wrong idea.[4]

Or think of all those history flicks which feature women with hourglass figures. Think that's how women used to look? Sorry, they did not. Through most of history women did not come wide in the chest and tight in the middle. Most of the time they just came wide.*

Another area in which the moviemakers have had a little trouble is in geography. Costume movies are forever being shot in places where the action never actually happened. The Egyptian scenes in the 1963 version of *Cleopatra* were shot in Britain and Rome. They hadn't intended to shoot in Rome at all. But with the onset of the cold weather in London they had to. It just didn't seem right when the actors began speaking their lines under the hot Egyptian sun and cold steamy vapors came out.

The mistake Cecil B. DeMille made in filming the famous scene where Moses parts the Red Sea is that he reinforced the belief that the event took place at the Red Sea. It did not. It took place, according to modern scholars, at the Sea of Reeds, a marshy area located in northern Egypt. A mistranslation is responsible for the error.**

Sometimes the trouble with shooting a movie where it should be shot is that the authentic location simply doesn't seem as authentic as it should. Thus, in one of the most famous movies ever made about Jesus Christ, *The Greatest Story Ever Told* (1965), the action was filmed in Utah, not the Mid-

---

* In the 1963 remake of *Cleopatra*, we are reliably informed, "the average measurements of the one hundred extras playing handmaidens, palace servants, and priestesses were 37-24-36."

** In 1976 Burt Lancaster starred in a movie about Moses accurately depicting the crossing at the Sea of Reeds. The movie flopped.

dle East. Director George Stevens explained that the mesas in Utah looked realer than the real thing.

Speaking of Jesus, he didn't die the way the movies say he did. He did not carry the cross with him when he went up the hill to die. He only carried the cross-beam. The post was already in the ground.[5]

SOLDIERS AND WAR

In the movies about war a lot of soldiers usually die. Just as in real life. But they usually don't seem to suffer as much as they should.

There are other problems. Take Hollywood's idea of the Roman army. The movies always show Roman soldiers marching, but Roman soldiers didn't march. Historians tells us nobody marched to war until the eighteenth century, when the Prussians invented marching.[6]

The Roman navy the movies get wrong as well. You know how they always show slaves rowing in the galleys. Well, only free men were allowed to row. Slaves were considered too undependable.

Where did moviemakers get the idea that slaves did the rowing? They got it from a historian, and one of the greatest historians at that: Theodore Mommsen. Unfortunately, as has now been established, Mommsen made a mistake.

The Romans, being legalistically minded and all, even had a law against employing slaves in the galleys. Thus, say, in a pinch, an admiral in absolutely desperate circumstances needed to borrow somebody's slaves for galley rowing, as sometimes happened—he would first give them their freedom. Only then would he throw them into the hole and make them start rowing.[7]

The movies do Roman gladiators pretty well. But they don't kill off nearly as many as they should. From the movies, for example, people have gotten the idea that only a few dozen or so gladiators would get killed at a typical event at the Colisseum. But in fact thousands would sometimes die. On one occasion something like 3,000 gladiators were made to hack each other to death.[8]

Speaking of Greeks reminds me of Hercules. You know what the problem with the Hercules movies is? They always portray Hercules as a knight in shining armor without the armor. When he wrestles the lion to the ground it is always before a crowd of dignitaries seated in boxes, as if he were at some kind of medieval tournament or something. And almost always in the movies there comes a point where he rescues some girl, as if she were a medieval damsel in distress and he was some kind of castle knight. The fact is, real Herecules types wouldn't have behaved the same way: they weren't gentlemen.

Why then do the movies portray them as if they were? Because the movies borrowed the image of Hercules from medieval romantics, and the romantics pretended Hercules was some kind of romantic medieval hero.[9]

NEWSREELS

If the movies often get history wrong, it's thought that the newsreels—many of which were also produced by Hollywood—got history-in-the making right. At the very least the newsreels always offered splendid pictures, however slanted or sentimental the narration may have been. But the pictures often were bogus. Unbeknownst to movie audiences—or to film historians for many years, for that matter—the pictures were often staged or faked outright.

Royal ceremonies, for example, were sometimes faked for the benefit of the cameras. Newsreel footage of the coronation of King Edward VII was shot several months before the event took place. This was accomplished through the help of actors. The actual ceremony had to be delayed because of the illness of the king. The fake one, however, occurred precisely on schedule.

Phony war scenes were especially common. During the Spanish-American War audiences were shown an exciting moment purportedly of the Battle of Santiago, which was actually shot in a bathtub containing toy ships. An astonished witness to the re-creation, commenting on its sophistication, reported that "electrically controlled devices supplied waves, and push buttons controlled the guns and ship movements."

Newsreel footage supposedly of the Boer War was also faked. Film historians have now established that much of it was shot in New Jersey. It was shot there because the company responsible for making the newsreels was located in New Jersey. The company was run by Thomas Alva Edison.

Footage of World War I was often staged, sometimes with the cooperation of enemy German soldiers. When photographer Donald Thompson needed action footage of the German Ninth Army, he simply asked the Germans for help and they gave it, destroying a windmill.

The reason for faking the pictures of the war was that the fake pictures always looked more authentic than the real ones. Captain F. E. Kleinschmidt, who travelled with the Austrian army, explained: "In real life a man who has been hit by a bullet does not throw up his hands and rifle and then fall in a theatrical fashion and roll a few times over. When he lies in the trenches and is hit he barely lurches a few inches forward or quietly turns on his side. The real picture is not as dramatic as the fake picture."

D. W. Griffith, on a visit to the front to obtain footage for *Hearts of the World,* found that even major battles did not seem theatrical on film. This was partly because he could not set up his camera in the no man's land between the two sides, where the view was best. But it was also because the most dramatic fighting often took place at night. As a cameraman with the U.S. Signal Corps reported, "when conditions are good for fighting they are, of necessity, poor for photography."

Actual footage of World War I proved so disappointing that the American committee in charge of war propaganda, the Committee on Public Information, sent an employee over to Europe to find out the reason. He reported back that the problem was that the battles were waged over too broad a front: "If you take a wide range of a battle going on all you get is a lot of shells bursting. There's no way of showing hundreds of men taking a charge because they don't go forward in close formation. You're lucky if you can get half a dozen figures in the range of your camera."[10]

In the Spanish Civil War, we are now told, British newsreels included numerous deceptions, including an aerial dogfight, left over from World War I. In November 1936, Gaumont British News ran a clip supposedly showing "the fall of Madrid," though the city did not actually fall until two and a half years later. The footage audiences saw was of the city of Burgos, 132 miles north of Madrid.

Even the much-respected "March of Time" newsreels contained fake footage. A 1937 newsreel, for instance, featuring the Japanese attack on China was shot in the United States.* In 1940, newsreel photographers staged a Japanese award ceremony in which an American captain was honored for saving the lives of several Japanese sailors. Newsreel photographers decided when the ceremony

---

* Yes, in New Jersey.

would take place, when the participants should start and stop talking, and what they should say. The first time the ceremony was stopped was to make everybody quiet down. The second time was to instruct the captain to turn to the camera when he came to a certain line. The third time was to have the captain repeat his speech so the cameraman could shoot a close-up. The fourth time was to tell the captain to change his speech.* [11]

In World War II the newsreels showed Hitler doing a jig after the fall of France in 1940. Millions saw him dance, but he never did. The illusion of the jig was created by trick photography. All Hitler actually did was raise his leg. But with the help of something called an optical printer, he was made to look as if he were a raving maniac dancer.[12]

---

* To which request the captain responded: "You want the truth or a story?"

CONCLUSIONS

"History is always written wrong, and so always needs to be rewritten."

—George Santayana

"Though God cannot alter the past, historians can."

—Samuel Butler

"While the mediocre European is obsessed with history, the mediocre American is ignorant of it."

—Anonymous

"I often think it odd that [history] should be so dull, for a great deal of it must be invention."

—Catherine Morland (in Jane Austen's *Northanger Abbey*)

"Historians, it is said, fall into one of three categories: Those who lie. Those who are mistaken. Those who do not know."

—Anonymous

"I know histhry isn't thrue, Hinnessey, because it ain't like what I see ivery day in Halstead Street."

—Mr. Dooley (Finley Peter Dunne)

ACKNOWLEDGMENTS

When you write a book like this one you always want to make sure, since you're pointing out everybody else's errors, that you don't make any yourself. I want to assure readers that there's not one error in this book made on purpose. If any crept in it was an accident.

To help me catch errors, Craig Conant, an able and brilliant scholar with a broad background in history, reviewed the manuscript in detail.

Bernie Weisberger is now and always has been my mentor. I think when I was picking a mentor I picked pretty well.

Stephen McAdoo is not an editor by profession, but he might as well be. His careful dissection of the manuscript was acute.

Jeff Bernstein kindly helped me with the science section.

Richard Bartone helped by making available back issues of *Film and History*.

Cynthia Barrett, my editor at HarperCollins,

gave me the encouragement I needed to see the book through to conclusion. Because of her intelligent editing this is a much better book than it otherwise would have been. I am grateful for her help and her friendship.

Without the help of Ed Victor I would be out reporting news stories instead of writing history. For making the life of a writer possible, I am truly thankful.

If there is one person on whom I have relied more than any other all these years, it is Michael Reed.

NOTES

PART 1: WAY BACK WHEN

1. FitzRoy Raglan, *The Hero* (1956; rpt. 1975), pp. 98–108, 151–54, 159–72, 225; Glyn Daniel, ed., *Myth or Legend?* (1955), chapter 2.

2. J. B. Bury, *Selected Essays* (1930), chapter 6; William Ober, *Boswell's Clap and Other Essays* (1979), chapter 10.

3. George Woodcock, "Legendary Alexander," *History Today* (November 1970), pp. 762–70; N. G. L. Hammond, *Alexander the Great* (1980); Simon Hornblower, "Lived Fast, Died Young," *New York Times Book Review,* September 22, 1991, p. 54; Mary Renault, *The Nature of Alexander* (1975); Peter Green, *Alexander of Macedon* (1974), pp. 141, 407.

4. William Nelson, *Fact or Fiction: The Dilemma of the Renaissance Storyteller* (1973), pp. 4–8, 27, 77–78.

5. Bergen Evans, *The Spoor of Spooks* (1954), p. 52 says Caesar could not have been born through a cesarean; Ashley Montagu and Edward Darling, *The Prevalence of Nonsense* (1967), pp. 194–96 say Caesar could have been born through a cesarean. Tom Burnam agrees in *More Misinformation* (1980), pp. 44–46.

6. Zvi Yavetz, "Caesar, Caesarism, and the Historians," *Journal*

of Contemporary History, VI, No. 2 (1971), 184–95; R. A. G. Carson, "The Ides of March," *History Today* (March 1957), pp. 142, 145; C. E. Stevens, "Crossing the Rubicon," *History Today* (June 1952), pp. 373–78.

7. Lucy Hughes-Hallett, *Cleopatra: Histories, Dreams and Distortions* (1990); Michael Grant, "Cleopatra," *History Today* (August 1971), pp. 533–41; Tristram Potter Coffin, *The Female Hero in Folklore and Legend* (1975), chapter 2.

8. *New York Times,* 1 April, 1990, "Week in Review," p. 4; Ronald Mellor, ed., *From Augustus to Nero* (1990), chapter 9.

9. Mellor, *Augustus to Nero,* chapter 12; Michael Grant, "Nero: The Two Versions," *History Today* (May 1954), pp. 319–25; B. H. Warmington, *Nero* (1969), pp. 123–71.

10. C. E. Stevens, "The End of the Roman Empire," *History Today* (June 1955), p. 401.

11. Richard Haywood, *The Myth of Rome's Fall* (1958); Richard Hodges and David Whitehouse, *Mohammed, Charlemagne & the Origins of Europe* (1983); J. J. Saunders, "The Debate on the Fall of Rome," *History* (February 1963), pp. 1–17; William C. Bark, *Origins of the Medieval World* (1960), pp. 3, 10–11.

12. E. L. Woodward, *History of England* (1962), p. 3.

13. George M. Trevelyan, *An Autobiography and Other Essays* (1949), p. 90.

14. E. R. Chamberlin, "The Death and Resurrection of Rome," *History Today* (May 1978), pp. 304–312.

15. Haywood, *Myth of Rome's Fall,* pp. 98–101; Alfons Dopsch, *The Economic and Social Foundations of European Civilization* (1937), pp. 2, 89–92; Michael Grant, "Attila the Hun," *History Today* (May 1954), pp. 170–71.

PART 2: THE DARK AGES

1. Theodor Mommsen, *Medieval and Renaissance Studies,* ed., Eugene Rice (1959), pp. 106–07.

2. Norman Cantor, *Inventing the Middle Ages* (1991), p. 23.

3. Lawrence Stone, *The Past and the Present* (1981), p. 160.

4. Bergen Evans, *The Spoor of Spooks* (1954), p. 38.

5. J. H. Plumb, *The Making of an Historian* (1988), p. 367.

6. Kirkpatrick Sale, *The Conquest of Paradise* (1990), pp. 85–86.

7. J. H. Hexter, *Reappraisals in History* (1961), pp. 3–4; Vaclav Mudroch and G. S. Couse, eds., *Essays on the Reconstruction of Medieval History* (1974), 115–27.

8. Joseph Levine, *Humanism and History* (1987), pp. 85–86; Rosemary Jann, *The Art and Science of Victorian History* (1985), p. xxvii; Jeffrey B. Russell, *Inventing the Flat Earth* (1991), p. 65; Geoffrey Barraclough, *History in a Changing World* (1956).

9. Steven Runciman, *A History of the Crusades* (1951–1954); Shlomo Eidelberg, *The Jews and the Crusaders* (1977); Bernard McGinn, "The Piety of the First Crusaders," in *Essays on Medieval Civilization,* eds. Bede Lackner and Kenneth Philip (1978), pp. 42–43, 48, 50, 69.

10. John Barnie, *War in Mediaeval Society* (1974), pp. 67–95; Mark Girouard, *The Return to Camelot* (1981), pp. 16–27; A. Dwight Culler, *The Victorian Mirror of History* (1985), pp. 29, 152–55; Ruth Cline, "The Influence of Romances on Tournaments of the Middle Ages," *Speculum* (April 1945), pp. 204–09; Frances Gies, *The Knight in History* (1984); Sidney Painter, *King John* (1949), p. 353; C. Warren Hollister, "The Irony of English Feudal ism," *Journal of British Studies* (May 1963), pp. 4, 21; J. H. Plumb, *Death of the Past* (1970), p. 48; H. H. Leonard, "Distraint of Knighthood: The Last Phase, 1625–41," *History* (February 1978), pp. 23–25, 34–37.

11. Joseph Shatzmiller, *Shylock Reconsidered: Jews, Moneylending, and Medieval Society* (1990).

12. Henry Kamen, *The Spanish Inquisition* (1965); Edward Peters, *Inquisition* (1988).

PART 3: A NEW DAY DAWNS

1. Paul T. Durbin, ed., *A Guide to the Culture of Science, Technology, and Medicine* (1980), pp. 26–28.

2. Christopher Hill, *The Collected Essays* (1986), III, 280.

3. Hill, *Essays,* III, chapter 13; T. G. Ashplant and Adrian Wilson, "Present-Centered History and the Problem of Historical Knowledge," *Historical Journal,* XXXI, No. 2 (1988), 259–60; David Park, *The How and the Why: An Essay on the Origins and Development of Physical Theory* (1988), pp. 196–201; M. L. Righini Bonelli and William Shea, eds., *Reason, Experiment, and Mysticism in the Scientific Revolution* (1975); John G. Burke, ed., *Science & Culture in the Western Tradition* (1987), chapter 5.

4. I. B. Cohen, *Revolution in Science* (1985), chapter 7.

5. Cohen, *Revolution in Science,* pp. 140–41.

6. Harvey Einbinder, *The Myth of the Britannica* (1964), pp. 255–56.

7. Burke, ed., *Science & Culture in the Western Tradition,* p. 92.

8. Reuben Parsons, *Some Lies and Errors of History* (6th ed; 1893), p. 82.

9. Richard Westfall, "Newton and the Fudge Factor," *Science,* February 23, 1973, pp. 751–58; "Numbers That Lied," *New York Times,* January 28, 1990, section E, p. 7; Curt Stern and Eva Sherwood, eds., *The Origin of Genetics* (1966), Part Five; Cristine Russell, "In the Light of History, Pasteur Is Tarnished," *Washington Post National Weekly,* March 1–7, 1993, p. 38.

PART 4: THE FACTS OF LIFE

1. Unless otherwise indicated, information in this part is based on Lawrence Stone's *The Family, Sex and Marriage in England, 1500–1800* (1977) and Christopher Hill's *The Collected Essays,* III, chapter 9.

2. John Gillis, "Married but Not Churched," in *'Tis Nature's Fault,* ed. Robert Maccubbin (1985), pp. 34–35.

3. G. S. Rousseau, "The Pursuit of Homosexuality in the Eighteenth Century," in *'Tis Nature's Fault,* ed. Maccubbin, pp. 152, 159.

4. Iwan Bloch, *Marquis de Sade* (1931); Gert Hekman, "Rewriting the History of Sade," *Journal of the History of Sexuality* (1990), pp. 131–36.

5. Michael Delon, "The Priest, the Philosopher, and Homosexuality in Enlightenment France," in *'Tis Nature's Fault,* ed. Maccubbin, p. 123.

6. Christina M. Root, "History as Character," in *History and Myth,* ed. Stephen Behrendt (1990), p. 151.

7. Vern Bullough, "Prostitution and Reform in Eighteenth Century England," in *'Tis Nature's Fault,* ed. Maccubbin, pp. 60–72; Roy Porter, "Mixed Feelings," in *Sexuality in Eighteenth-Century Britain,* ed. Paul-Gabriel Bouce (1982), pp. 8–9.

8. Harvey Einbinder, *The Myth of the Britannica* (1964), p. 94.

9. Charles Seltman, "Diogenes," *History Today* (February 1956).

10. Einbinder, *Britannica,* pp. 118–20; J. H. Plumb, *The Making of an Historian* (1988), pp. 348–51.

11. Plumb, *Historian,* pp. 348–49.

12. David Kertzer, "Gender Ideology and Infant Abandonment in Nineteenth-Century Italy," *Journal of Interdisciplinary Studies* (Summer 1991), pp. 1–25; William Langer, "Infanticide: A Historical Survey," in *The New Psychohistory,* ed. Lloyd deMause (1975),

chapter 3; Lawrence Stone, *The Past and the Present* (1981), pp. 216–18; Rene Leboutte, "Offense Against Family Order," *Journal of the History of Sexuality* (October 1991), pp. 159–85; David Brion Davis, *From Homicide to Slavery* (1986), p. 169.

13. John Boswell, *Christianity, Social Tolerance, and Homosexuality* (1980), pp. 100, 102, 135, 171, 180, 205–06, 213, 298, 333.

14. James Steakley, *The Homosexual Emancipation in Germany* (1975), pp. 103–21.

PART 5: GOD SAVE THE KING!

1. Eric Hobsbawm, "Inventing Tradition," and David Cannadine, "The British Monarchy and the Invention of Tradition," in *The Invention of Tradition,* eds. Hobsbawm and Terence Ranger (1984), chapters 1, 4; Valerie Chancellor, *History for Their Masters* (1970), chapter 2; Alan Lloyd, *The King Who Lost America* (1971), chapter 1.

2. Hobsbawm, *Invention of Tradition,* chapters 1, 4.

3. Steven Runciman, "Richard Coeur-de-Lion," *History Today* (April 1955), pp. 219–27; Robert Birley, *The Undergrowth of History* (1969), pp. 22–23.

4. Lord Raglan, *The Hero* (1975), pp. 206–13.

5. A. J. Pollard, *Richard III* (1991); A. R. Myers, "The Character of Richard III," *History Today* (August 1954), pp. 511–21; Myers, "Richard III and Historical Tradition," *History* (June 1968), pp. 181–202.

6. Ian Christie, "George III and the Historians—Thirty Years On," *History* (June 1986), pp. 205–10; J. H. Plumb, *The American Experience* (1989), pp. 50–60; Richard Schlatter, ed., *Recent Views on British History* (1984), pp. 201–02.

7. David Cannadine, "The Last Hanoverian Sovereign?" in *The First Modern Society,* ed. A. L. Beier et al. (1989), pp. 127–65; Hobsbawm, *Invention of Tradition,* pp. 133–34; Dorothy Thompson, *Queen Victoria* (1990), chapter 4.

8. Philip Ziegler, *Edward VIII* (1992); Sarah Bradford, *The Reluctant King* (1989); Fulton Oursler, Jr., "Secret Treason," *American Heritage* (December 1991), pp. 52–68.

PART 6: "THIS SCEPTER'D ISLE"

1. A. F. Pollard, "Magna Carta," *History* (October 1917), pp. 170–73; W. L. Warren, "What Was Wrong with King John?" *History Today* (December 1957), pp. 806–11; C. G. Crump, "The Execution of the Great Charter," *History* (October 1928), pp. 247–53; Edward P. Cheyney, "The Disappearance of English Serfdom," *English Historical Review*, XV (1900), 20–37; G. T. Hankin, "Magna Carta in the U.S.A.," *History* (March 1940), pp. 318–21; Warren C. Hollister, "King John and the Historians," *Journal of British Studies* (November 1961), pp. 1–19; J. C. Holt, *King John* (1963).

2. G. R. Elton, *Star Chamber Stories* (1958), pp. 11–12; Elton, *The Tudor Constitution* (2d ed.; 1982), chapter 6; J. P. Kenyon, ed., *The Stuart Constitution* (1986), pp. 104–07; A. L. Rowse, *The England of Elizabeth* (1951), p. 364.

3. Lawrence Stone, "The Armada Campaign of 1588," *History* (September 1944), pp. 120–43; Cynthia F. Behrman, *Victorian Myths of the Sea* (1977), pp. 78–86; Geoffrey Callender, "The Real Significance of the Armada's Overthrow," *History* (October 1917), 174–77; Garrett Mattingly, *The Armada* (1959); Geoffrey Parker, "Why the Armada Failed," *History Today* (May 1988), pp. 26–33.

4. Alexander Winston, "Captain Kidd," *History Today* (October 1965), pp. 727–33.

5. Against the story is: Bergen Evans, *The Spoor of Spooks* (1954), pp. 69–74; in favor of it is: Ashley Montagu and Edward Darling, *The Ignorance of Certainty* (1970), pp. 120–25.

6. Percy Adams, *Travelers and Travel Liars* (1980), pp. 172–75, 261.

7. Behrman, *Victorian Myths of the Sea*, chapter 7.

8. Phillip Knightley and Colin Simpson, *The Secret Lives of*

Lawrence of Arabia (1969); John Mack, *A Prince of Our Disorder* (1976), pp. 222–33, 422–41; Malcolm Brown, ed., *T. E. Lawrence: The Selected Letters* (1989), General Introduction.

9. Eric Hobsbawm, "Mass-Producing Traditions," in *The Invention of Tradition,* eds. Hobsbawm and Terence Ranger (1984), p. 295.

10. Herbert Butterfield, *The Englishman and His History* (1944; rpt. 1970), pp. 75–76.

11. Hugh Trevor-Roper, "The Invention of Tradition: The Highland Tradition of Scotland," in *Invention of Tradition,* pp. 15–41.

PART 7: LET THEM EAT BRIOCHE!

1. C. S. B. Buckland, "Saint Joan," *History* (January 1925), pp. 273–81; Marc Ferro, *The Use and Abuse of History* (1984), pp. 105–112; Charles W. Lightbody, *The Judgments of Joan* (1961).

2. Reuben Parsons, *Some Lies and Errors of History* (6th ed; 1893), pp. 113–20; Bergen Evans, ed., *Dictionary of Quotations* (1968), p. 658.

3. Evans, *Dictionary,* p. 85.

4. Eileen Simpson, *Orphans* (1987), pp. 158–63; Albert Schinz, Review of *The Political Writings of Jean Jacques Rousseau,* ed. C. E. Vaughan, *Philosophical Review* (January 1917), pp. 214–27; Percy Adams, *Travelers and Travel Liars* (1980), pp. 125–26, 197.

5. Ashley Montagu and Edward Darling, *The Ignorance of Certainty* (1970), pp. 96–97; Norbert Guterman, ed., *A Book of French Quotations* (1963), p. 188; Theodore Besterman, ed., *Voltaire's Correspondence* (1962), LXXIV, 80.

6. Esmond Wright, "Lafayette: Hero of Two Worlds," *History Today* (October 1957), pp. 656–61.

7. Arno Karlen, *Napoleon's Glands* (1984), chapter 1; Pieter Geyl, *Debates with Historians* (1955), chapter 11; Vincent Cronin, *Napoleon* (1990).

8. David Chandler, *The Campaigns of Napoleon* (1966), pp. 852–5; E. H. Dance, *History the Betrayer* (1964), p. 21; Cronin, *Napoleon,* chapters 21, 25; A. J. P. Taylor, *Europe: Grandeur and Decline* (1967), p. 12.

9. Douglas Johnson, "L'Affaire," *History Today* (July 1935), p. 5.

PART 8: LIKEABLE (AND NOT-SO-LIKEABLE) FAMOUS PEOPLE

1. Allan Nevins, "Machiavelli Not So Machiavellian," *New York Times Magazine* (6 October 1940), p. 6ff.; H. R. Trevor-Roper, *Men and Events* (1957), chapter 8; Simon Harcourt-Smith, "Machiavelli," *History Today* (January 1956), pp. 45–53; Barrows Dunham, *Man Against Myth* (1947), pp. 214–21.

2. A. Lentin, "Catherine the Great and Enlightened Despotism," *History Today* (March 1971), pp. 170–77.

3. Sterling Seagrave, *The Soong Dynasty* (1985); Barbara Tuchman, *Sand Against the Wind: Stillwell and the American Experience* (1971), pp. 40–45, 50–51, 115–16.

4. Seagrave, *Soong Dynasty;* Tuchman, *Sand Against the Wind,* pp. 116, 132; Jonathan D. Spence, *The Search for Modern China* (1990), p. 402.

5. Richard Grenier, "The Gandhi Nobody Knows," *Commentary* (March 1983), pp. 59–72; William Shirer, *Gandhi: A Memoir* (1979).

PART 9: KING ARTHUR AND SUCH

1. FitzRoy Raglan, *The Hero: A Study in Tradition, Myth, and Drama* (1975), chapter 6; Glyn Daniel et al., *Myth or Legend* (1955), chapter 3; Janet and Colin Bard, *Ancient Mysteries of Britain* (1986), pp. 168–81; T. F. Tout, *Edward the First* (1903), p. 117; Esmé Wingfield-Stratford, *Truth in Masquerade* (1951), p. 90; Christina Hole, *English Folk-Heroes* (1948), pp. 38–56; Frances

Gies, *The Knight in History* (1984), p. 2; Valerie Logario and Mildred Day, eds., *King Arthur Through the Ages* (1990), II, 154–57; J. H. Plumb, *The Making of an Historian* (1988), p. 281.

2. Hole, *English Folk-Heroes,* pp. 15–17.

3. Hole, *English Folk-Heroes,* chapters 5, 6; Maurice Keen, "Robin Hood: A Peasant Hero," *History Today* (October 1958), pp. 684–89; Keen, *The Outlaws of Medieval Legend* (1977); Raglan, *Hero,* chapter 4.

4. Wolfgang Mieder, *Tradition and Innovation in Folk Literature* (1987), chapter 2.

5. Alan Dundes, lecture at the University of California at Berkeley, September 9, 1991.

6. Paul Dukes, "Dracula: Fact, Legend and Fiction," *History Today* (July 1982), pp. 44–47.

7. Chris Baldick, *In Frankenstein's Shadow* (1987), pp. 199–204.

8. Bergen Evans, *The Spoor of Spooks* (1954), pp. 60–61.

9. Charles Jones, *Saint Nicholas* (1978); Jones, "Knickerbocker Santa Claus," *New-York Historical Society Quarterly* (October 1954), pp. 357–83; Eric Wolf, "Santa Claus: Notes on a Collective Representation," in *Process and Pattern in Culture,* ed. Robert Manners (1964), pp. 147–55.

PART 10: RELIGION

1. Unless otherwise indicated material in Part 10 is drawn primarily from Robin Lane Fox's *The Unauthorized Version: Truth and Fiction in the Bible* (1992).

2. John Boswell, *Christianity, Social Tolerance, and Homosexuality* (1980), pp. 93–97.

3. H. R. Trevor-Roper, "The Uses of Fakery," *New York Review of Books* (December 6, 1990), p. 28.

4. S. G. F. Brandon, "The Jesus of History," *History Today* (January 1962), p. 14.

5. Nathan H. Dole, *The Mistakes We Make* (1898), p. 100.

6. J. K. Elliott, "The Birth and Background of Jesus of Nazareth," *History Today* (December 1978); S. G. F. Brandon, "The Jesus of History," *History Today* (January 1962); Peter Steinfels, "Historical Jesus Ever Elusive," *International Herald Tribune*, December 24–25, 1991, p. 1ff.

7. Richard Haywood, *The Myth of Rome's Fall* (1958), p. 92; J. H. Plumb, *The Making of an Historian* (1988), p. 366; R. A. G. Carson, "The Emperor Constantine and Christianity," *History Today* (January 1956), pp. 18–19.

8. Peter Gay and R. K. Webb, *Modern Europe* (1973), pp. 156–57.

9. Lawrence Stone, *The Family, Sex and Marriage in England, 1500–1800* (1977), pp. 42–43.

10. Garry Wills, *Under God* (1990), p. 334.

11. G. G. Coulton, *Studies in Medieval Thought* (1940), p. 78.

PART 11: WORLD WARS I AND II

1. Thomas A. Bailey, *A Diplomatic History of the American People* (8th ed.; 1969), p. 24; A. J. P. Taylor, *Europe: Grandeur and Decline* (1967), pp. 183–89; Herbert Butterfield, *George III and the Historians* (1959), pp. 32–34; Anthony Kemp, *The Maginot Line: Myth and Reality* (1981).

2. Richard S. Geehr, "The Last Nazi," *Film and History* (May 1979), p. 44.

3. A. J. P. Taylor, "Who Burnt the Reichstag?" *History Today* (August 1960), pp. 515–22.

4. E. H. Dance, *History the Betrayer* (1964), p. 61.

5. Michael Howard, *The Lessons of History* (1991), pp. 49–62; E. J. Feuchtwanger, *Prussia: Myth and Reality* (1970), pp. 9–11, 54–59.

6. V. R. Berghahn, "The Twisted Road to Auschwitz," *New York Times Book Review*, February 19, 1989, p. 28; Dick Geary, "Image

and Reality in Hitler's Germany," *European History Quarterly,* XIX (1989), 385–87; A. J. P. Taylor, *A Personal History* (1984), p. 299; John Kenneth Galbraith, *A Life in Our Times* (1981), pp. 196–214.

7. William Carr, "A Final Solution? Nazi Policy Towards the Jews," *History Today* (November 1985), pp. 30–36. For a contrary view see Daniel Goldhagen, "The Road to Death," *New Republic* (November 4, 1991), p. 34ff.

8. Paul Kennedy, "Appeasement," *History Today* (October 1982), pp. 51–53; D. C. Watt, "The Rise and Fall of the Third Reich," *History* (June 1970), pp. 215–16; Clive Ponting, *1940* (1990); John Lukacs, "The Transatlantic Duel: Hitler vs. Roosevelt," *American Heritage* (December 1991), p. 71; Frederick Marks, III, "Six Between Roosevelt and Hitler," *Historical Journal,* XXXVIII, No. 4 (1985), 969–82.

9. William Shirer, *A Native's Return* (1990), pp. 438–39; Shirer, *The Collapse of the Third Republic* (1969), pp. 611–20.

10. A. J. P. Taylor, *Politicians, Socialism, and Historians* (1982), pp. 230–35; Ponting, *1940.*

11. Stanley Weintraub, "Three Myths about Pearl Harbor," *New York Times,* December 4, 1991, p. A–19.

12. Stephen Ambrose, "Ike and the Disappearing Atrocities," *New York Times Book Review,* February 24, 1991, p. 1ff.

13. Glenn B. Infield, *Hitler's Secret Life* (1979), chapter 6.

14. Ian Kershaw, "The Hitler Myth," *History Today* (November 1985), pp. 28–29.

15. Denis Mack Smith, "Mussolini, Artist in Propaganda," *History Today* (April 1959), pp. 223–32.

16. David F. Schmitz, *The United States and Fascist Italy* (1988).

17. Ponting, *1940,* especially chapter 6; Henry B. Ryan, "A New Look at Churchill's 'Iron Curtain' Speech," *Historical Jour-*

nal, XXII (1979), 897–98; *American Heritage* (February-March 1991), p. 30.

18. Edward Behr, *Hirohito* (1990).

PART 12: HOLLYWOOD DOES HISTORY

1. Information in this section concerning hairstyles and costumes, unless otherwise indicated, comes from Edward Maeder, ed., *Hollywood and History: Costume Design in Film* (1987).

2. Quentin Bell, "Dressing the Past," *History Today* (July 1951), pp. 44–45.

3. George Fraser, *The Hollywood History of the World* (1988), pp. xv–xvi.

4. Derek Elley, *The Epic Film* (1984), p. 53.

5. Jon Solomon, *The Ancient World* (1978), pp. 97–98.

6. Denys Arcaud, "The Historical Film," *Cultures,* VII, No. 1 (1974), 24.

7. Chester Starr, *The Roman Imperial Navy* (1960), pp. 66–74.

8. Keith Hopkins, "Murderous Games," *History Today* (June 1983), pp. 16–22.

9. Joseph Levine, *Humanism and History* (1987), pp. 20–21.

10. David Mould and Charles Berg, "Fact and Fantasy in the Films of World War I," *Film and History* (September 1984), pp. 50–59.

11. John O'Connor, ed., *Image as Artifact* (1990), p. 73; Anthony Aldgate, "British Newsreels and the Spanish Civil War," *History* (February 1973), pp. 60–63; Raymond Fielding, *The American Newsreel* (1972), chapter 15; Paul Smith, ed., *The Historian and Film* (1976), pp. 100, 118.

12. O'Connor, *Image as Artifact,* p. 11.

INDEX

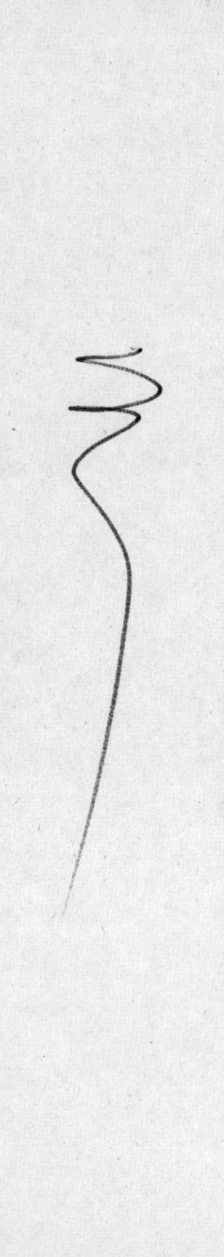